Qualitative Methods in Physical Kinetics and Hydrodynamics

Qualitative Methods in Physical Kinetics and Hydrodynamics

Vladimir P. Krainov

Translated by
Kevin Hendzel

Library of Congress Cataloging-in-Publication Data

Kraĭnov. V. P. (Vladimir Pavlovich), 1938
[Kachestvennye metody v fizicheskoĭ kinetike i gidrogazodinamike. English]
Qualitative methods in physical kinetics and hydrodynamics / Vladimir P. Krainov; translated by Kevin Hendzel.
p. cm.--
Translation of: Kachestvennye metody v fizicheskoĭ kinetike i gidrogazodinamike.
Includes bibliographical references and index.
ISBN 978-0-88318-953-5
1. Gas dynamics. 2. Hydrodynamics. 3. Mechanics, Analytic. I. Title. II. Series:
QC168.2.K73 1992
532'.05—dc20 91-25909
CIP

Contents

Preface

This book, which is based on a course of lectures given by the author at the Moscow Engineering-Physics Institute, can be used as a supplement to existing texts on physical kinetics and hydrodynamics. Its purpose is to deepen the student's understanding of the subject and to help the student master the basic elements of the physics of irreversible statistical processes.

The text contains solutions to the basic problems of physical kinetics, hydrodynamics, gas dynamics, and thermodynamics, worked out by qualitative methods. By the term "qualitative," we mean that all results are obtained to the correct order of magnitude. Numerical factors of order unity, the calculations of which require exact solution of the kinetic equations or hydrodynamic equations, are omitted. Qualitative methods are important because numerical calculations in physical kinetics and hydrodynamics are usually difficult and expensive in computer execution time. Before considering such calculations, it is advisable to estimate the results qualitatively. The qualitative method does not require the use of a computer or the analytical solution of the differential and integral equations of physical kinetics or hydrodynamics (analytical solutions are often either impossible or lead to complicated special functions of mathematical physics).

Qualitative estimates give the correct orders of magnitude of the physical quantities obtained in the solution of the problem and quite often can be used to simplify the equations themselves (which are then implemented in computer programs) by discarding unimportant terms in the equations. The unimportance of a term is revealed only through the use of qualitative approaches and is not obvious beforehand.

The differential equations of hydrodynamics and thermodynamics contain various kinetic characteristics of the medium as coefficients: the thermal conductivity, viscosity, diffusion coefficient, and so on. These quantities are estimated in the first few chapters of the present text. The purpose of the later chapters is to give qualitative solutions of the equations describing various macroscopic nonequilibrium processes in media. Here, the kinetic coefficients characterizing the medium are assumed to be given.

In the first few chapters we consider the kinetic properties of neutral gases, weakly ionized gases, plasmas, dielectrics, and metals. In the later chapters we consider nonequilibrium processes in ideal and viscous fluids, and the phenomena of turbulence, heat conduction, diffusion, and free convection in typical problems of hydrodynamics, gas dynamics, boundary-layer physics, and surface phenomena, and the propagation of various waves in media.

After reading and mastering the material of each section, it is recommended that the student obtain qualitative solutions to the problems given at the end of the section. Many of the problems have different methods of solution, which lead (obviously) to the same final result. The problems discussed in the text are usually solved using only one approach. It is recommended that the reader try other methods of solving the problem.

In physical kinetics and hydrodynamics there are obviously a large number of problems in which the qualitative approach to the solution is either trivial or impractical. If quantitative analytical solutions to these problems exist, then it is usually evident at once whether or not it is possible to obtain the general solution of the problem (or a solution in some limiting case) by qualitative methods. If an analytical solution does not exist, then qualitative approaches can be based on experimental information.

The material of the present text should be understood with a knowledge of undergraduate physics. As for mathematical knowledge, the student is required to be able to correctly estimate and approximate mathematical expressions. For example, if we have a function of the form $\epsilon(T) = (T + T_0)^{-1}$, then if $T \ll T_0$ the derivative of this function is approximately ϵ/T_0, while if $T \gg T_0$ the derivative is approximately ϵ/T. In the absence of an explicit mathematical expression the derivative can be approximated from a graph of the function. For example, in the estimate (5.3) for the electronic heat capacity of a metal $c = d\epsilon/dT$ (T is the temperature, ϵ is the energy of an electron) the result is T/ϵ, and not ϵ/T, as one might think at first glance.

The author sincerely thanks S. P. Goreslavskiĭ, V. S. Imshennik, E. E. Lovetskiĭ, and V. P. Silin for carefully reading the manuscript and offering a large number of valuable suggestions, which were taken into account in the completed version of the book.

Chapter 1

Neutral gases

This chapter considers nonequilibrium processes in rarefied neutral gases, i.e., gases consisting of neutral atoms or molecules.

What is a rarefied gas? Each gas molecule moves almost all of the time as a free particle and interacts with the other molecules only in rare collisions (however, the interactions in the collisions cannot be assumed to be weak). Hence we assume that the average distance between molecules $\langle r \rangle \sim n^{-1/3}$ (n is the concentration of molecules; the number of molecules in a volume of 1 m^3 of gas) is large in comparison with the range d of the intermolecular forces: $\langle r \rangle \gg d$ or $nd^3 \ll 1$. For collisions of neutral molecules with one another the quantity d is of the order of a molecular diameter ($d \sim 0.1$ nm). For a gas at atmospheric pressure $n \sim 10^{25}$ m^{-3} and therefore $\langle r \rangle \sim 1$ nm.

We consider the question of the use of classical mechanics or quantum mechanics to describe the excitation of the different degrees of freedom of the gas molecules. The translational motion of the molecules can always be treated classically. The rotation of molecules is also practically always classical, since the separation between neighboring rotational levels is of order $\hbar^2/I$, where $\hbar$ is Planck's constant and I is the moment of inertia of a molecule, and is small in comparison with ordinary temperatures T of a gas: $(\hbar^2/I) \ll T$ (below we will express the temperature T in energy units, such that the Boltzmann constant $k = 1$). Therefore if the temperature T is not very low, the thermal motion of the molecules excites rotational levels with large rotational quantum numbers and hence the rotation of the molecules is classical. This statement becomes incorrect at temperatures

$$\hbar^2/(Md^2) \sim (m/M)\text{Ry},$$

(where Ry = 27.2 eV), i.e., at temperatures of 1–10 K, since the mass M of a molecule is about 10^5 times larger than the mass m of the electron (here we have used the approximation $I \sim Md^2$ for the moment of inertia of a molecule). But at such low temperatures the gas has already transformed into a liquid or solid. (The relation 1 eV $\sim 1.1 \times 10^4$ K was used to obtain the temperatures discussed above.)

As for the vibrational motion of atoms in a molecule, under ordinary conditions (for temperatures which are not too high) vibrations will normally not be excited and the molecule will be in its ground vibrational state. The separation between neighboring vibrational levels is equal to $\hbar\omega$, where ω is the classical frequency of vibration of the atoms in the molecule. The energy $\hbar\omega$ corresponds to a temperature

$\hbar\omega \sim \sqrt{m/M}$ Ry $\sim 10^3$ K. This is the order of the temperature at which the vibrational levels begin to be excited.

1.1. Mean free path and mean free time

Kinetic processes in gases are determined by collisions between molecules. We first estimate the parameters describing these collisions.

Consider first the distance traveled by a molecule between two successive collisions, the so-called *mean free path* l. Let σ be the cross section for collisions of molecules with one another. Obviously in the case of neutral molecules, which can be approximated as hard spheres of diameter d, we have $\sigma \sim d^2$. Let a molecule (which does not move in a straight line because of collisions) travel a distance L. Over this distance it collides with all molecules inside a cylinder of length L and cross-sectional area σ. The number of such molecules is $n\sigma L$ (here n is the concentration of molecules). This number is then the number of collisions in the distance L. Therefore $l \sim L/(n\sigma L)$ or

$$l \sim \frac{1}{n\sigma}. \tag{1.1}$$

Setting $n \sim \langle r\rangle^{-3}$, where $\langle r\rangle$ is the average distance between molecules, and $\sigma \sim d^2$, we can write (1.1) in the form

$$l \sim \left(\frac{\langle r\rangle}{d}\right)^2 \langle r\rangle. \tag{1.2}$$

Since $\langle r\rangle \gg d$ in a rarefied gas, it follows from (1.2) that $l \gg \langle r\rangle$, i.e., the mean free path of the molecules in a rarefied gas is large compared to the average distance between molecules (and hence $l \gg d$). For a gas at room temperature and at atmospheric pressure we have $l \sim 100$ nm. We see that the above inequalities are satisfied, using typical values for $\langle r\rangle$ and d.

The *mean free time* τ is defined as $\tau = l/v$, where v is the typical velocity of a molecule, i.e., the average thermal velocity. From the Maxwell distribution we have

$$v \sim \sqrt{T/M}. \tag{1.3}$$

Here, M is the mass of a molecule. In particular, for air at room temperature (1.3) gives $v \sim 300$ m/s.

From (1.1) and (1.3) we obtain the following estimate for the mean free time:

$$\tau \sim \frac{1}{N\sigma}\sqrt{\frac{M}{T}}. \tag{1.4}$$

The quantity τ characterizes the time between two successive molecular collisions. Note that because of the condition $l \gg \langle r\rangle$, the velocities of randomly moving molecules before and after the collision are uncorrelated.

From the point of view of physical kinetics, the quantity τ plays the role of a *relaxation time* for the establishment of equilibrium in a volume element of the gas disturbed from the equilibrium state by an external perturbation, since equilibrium is attained by means of molecular collisions.

Because of the random motion of the molecules, the kinetic characteristics of a gas fluctuate. For example, the macroscopic flux of a gas is the average flux of a large number of molecules. The scatter of the true values of a quantity about the mean characterizes the dispersion. We will always consider the situation when such fluctuations are relatively small, i.e., the gas is satisfactorily described by the average physical quantities. This will only be true if the characteristic length L over which the macroscopic parameters vary (leading to a deviation from statistical equilibrium) is large in comparison with the average distance between molecules: $L \gg \langle r \rangle$. When this condition is satisfied a kinetic description of the system in terms of macroscopic kinetic characteristics is valid.

In the greater part of this chapter (except for Sec. 1.6) and in the succeeding chapters we will assume that the more restrictive condition $L \gg l$ is satisfied. In this case the microscopic structure of the gas disappears altogether and its properties can be described in terms of the macroscopic equations of hydrodynamics. These equations contain the kinetic coefficients (such as the diffusion coefficient, the viscosity, the thermal conductivity, and so on) as parameters, which in macroscopic hydrodynamics are treated as phenomenological parameters and are assumed given. Therefore *the purpose of physical kinetics is to calculate the kinetic coefficients in terms of the microscopic characteristics of collisions between individual molecules.* In the case $L \lesssim l$ the equations of hydrodynamics are not valid, but the calculation of the kinetic coefficients is still physically meaningful if the dispersions are small; this is discussed further in Sec. 1.6.

The condition $L \gg l$ defines a gas with slowly varying macroscopic parameters, which disturb the gas from the equilibrium state. We call such a gas a *slightly nonuniform* gas. A slightly nonuniform gas is described macroscopically by the equations of hydrodynamics, whereas the problem of kinetic theory is to calculate the kinetic coefficients appearing in these equations. Below we estimate the kinetic coefficients in the most typical situations.

Problem 1. A particle of mass m moves in a gas of molecules with mass M. Show that if $m \lesssim M$ the mean free path of the particle is given by (1.1), and if $m \gg M$ the mean free path is smaller by the factor $\sqrt{m/M}$.

Problem 2. Show that over a distance $x \ll l$ the probability of collision of a molecule is of order x/l.

1.2. Diffusion of one gas into another

As a first example of the determination of the kinetic coefficients we consider the diffusion of one gas into another. We will assume that the binary mixture is uniform in temperature and pressure.

We first consider the case when the concentration of one gas (gas 1) is small compared to that of the other gas (gas 2). We estimate the diffusion flux of gas 1 due to a concentration gradient dc/dx in gas 1. Here $c = n_1/n$, where n_1 is the concentration of molecules of gas 1, and $n = n_1 + n_2 \approx n_2$ is the total concentration of molecules, which is practically identical to the concentration n_2 of molecules of gas 2. Hence we assume that $c \ll 1$.

We have $dc/dx \sim c/L$, where L is the characteristic distance over which the concentration gradient of molecules of gas 1 is created. We assume that $L \gg l$, i.e., we assume that gas 1 is slightly nonuniform. Gas 2 is assumed to be uniform. The concentration gradient is assumed to be directed along a certain axis in space, which we call the x axis.

The number of molecules of the diffusing gas 1 passing through a unit area perpendicular to the x axis per unit time (the flux of the molecules) is equal to $n_1 v$. Consider the molecules of gas 1 flowing from left to right. Then the value n_1 should be calculated at a distance $x - l$ to the left of the reference area (x corresponds to the location of the reference area), since a molecule of gas 1 can reach the reference area from this position without collisions. Therefore the number of molecules under consideration is equal to $n_{1(x-l)}v$.

On the other hand, for the flux of molecules of gas 1 passing through the area from right to left we obtain the analogous expression $n_{1(l+x)}v$. The diffusion flux i is then equal to the difference of these two fluxes of molecules of gas 1:

$$i = [n_{1(x-l)} - n_{1(x+l)}]v \sim lv \frac{dn_1}{dx} = -lvn \frac{dc}{dx}. \tag{1.5}$$

In essence, (1.5) determines the variation in the number of particles as a result of collisions. When we expand (1.5) in a Taylor series and keep only the first derivative term, we assume that the quantity l is sufficiently small. The relation (1.5) is correct when gas 1 is slightly nonuniform, i.e., $l \ll L$. In this case we can neglect the higher-order terms of the expansion of the flux i in a Taylor series.

According to the phenomenological definition of hydrodynamics, we have

$$i = -D \frac{dn_1}{dx} = -nD \frac{dc}{dx}. \tag{1.6}$$

The quantity D is called the *diffusion coefficient*. Comparing (1.5) and (1.6), we obtain the estimate

$$D \sim lv. \tag{1.7}$$

We assume further that the mass M_1 of a molecule of gas 1 is small in comparison with the mass M_2 of a molecule of gas 2. Hence gas 1 will be called the light gas and gas 2 will be called the heavy gas, and we consider diffusion of a light gas into a heavy gas. We can assume that the heavy gas remains in a state of statistical equilibrium, since the nonrandom part of the velocity (along the x axis) acquired by heavy-gas molecules in collisions with the light molecules is small, in view of the large mass of the molecules of the heavy gas.

Substituting (1.1) and (1.3) into (1.7), we obtain an estimate of the diffusion coefficient of a light gas diffusing into a heavy gas:

$$D \sim \frac{1}{n\sigma} \sqrt{\frac{T}{M_1}}. \tag{1.8}$$

We write D in terms of the total pressure of the gas mixture $P = nT$ (which is determined by the heavy gas) according to the ideal-gas equation. We obtain

$$D \sim \frac{1}{P\sigma}\sqrt{\frac{T^3}{M_1}}. \tag{1.9}$$

Note that the diffusion coefficient does not depend on the concentration c of molecules of the diffusing gas.

Suppose a concentration gradient is created in the light gas and the gas is then left to itself. As a result of diffusion the gradient decreases with time and the concentration becomes uniform. We estimate the characteristic time τ_D of the diffusion process. We write the diffusion flux i of molecules of the light gas

$$i = n_1 V, \tag{1.10}$$

where V is the velocity of light molecules along the x axis induced by the concentration gradient. Comparing (1.10) with (1.6), we find

$$n_1 V = -D\frac{dn_1}{dx} \sim D\frac{n_1}{L}, \tag{1.11}$$

where L is the characteristic length over which the concentration difference was created, i.e., it is the characteristic scale of length of the problem. Hence we obtain for the induced velocity V

$$V \sim \frac{D}{L} \sim \frac{l}{L}v. \tag{1.12}$$

We have used (1.7) for the diffusion coefficient. We see that the induced velocity is small compared to the average thermal velocity of a molecule: $V \ll v$.

We define the diffusion time τ_D as the time required for a light molecule to travel a distance of the order of the characteristic length of the problem L, i.e., $\tau_D \sim L/V$. From (1.12) we find

$$\tau_D \sim \frac{L^2}{D} \sim \left(\frac{L}{l}\right)^2 \frac{l}{v} \gg \tau. \tag{1.13}$$

We see that *the diffusion time is large in comparison with the mean free time.*

A diffusion flux of the light gas can also be created by other means besides a concentration gradient. An example is a temperature gradient dT/dx (we assume the pressure in the gas mixture is uniform) created externally. In this case the phenomenon is called *thermal diffusion.* We define the thermal diffusion coefficient D_{td} according to the macroscopic hydrodynamic relation for the diffusion flux i:

$$i = -n\frac{D_{td}}{T}\frac{dT}{dx}. \tag{1.14}$$

The quantity D_{td} is defined in (1.14) such that it has the same dimension as the diffusion coefficient D. Because the concentration of the light gas is small by assumption, the gradient dT/dx is determined by the heavy gas. It can be considered as given, since the motion of the heavy molecules induced by the temperature gradient is negligible because of the large mass of the heavy molecules.

To determine the thermal diffusion flux of the light gas we proceed in analogy with the derivation of (1.5). However, in this case the average thermal velocity v of the light molecules varies with x, since it depends on the varying temperature according to (1.3). Therefore we obtain for the net diffusion flux i

$$i=n_1[v_{(x-l)}-v_{(x+l)}]\sim -ln_1\frac{dv}{dx}\sim -ln_1\frac{v}{T}\frac{dT}{dx}. \qquad (1.15)$$

In this relation the derivative dv/dT was approximated by v/T. Turning to (1.3), we see that this approximation introduces a numerical error (a factor of 2), but the dependence on the temperature T is correct.

Comparing (1.14) with (1.15), we obtain an estimate for the thermal diffusion coefficient D_{td}:

$$D_{td}\sim clv\sim cD\sim\frac{c}{n\sigma}\sqrt{\frac{T}{M_1}}\sim\frac{n_1}{p^2\sigma}\sqrt{\frac{T^5}{M_1}}. \qquad (1.16)$$

Here, we have used (1.8) and (1.9) for the diffusion coefficient. We see from (1.16) that unlike the diffusion coefficient, the *thermal diffusion coefficient depends on the concentration of molecules of the light gas.*

We consider now equilibrium in a nonuniform heated gas established as a result of diffusion. In diffusion equilibrium, the concentration distribution is such that the total diffusion flux is equal to zero. The diffusion flux due to the temperature gradient cancels out the diffusion flux in the opposite direction due to the concentration gradient in the light gas. We have

$$0=\frac{d}{dx}(n_1v)=\frac{d}{dx}(cnv)=P\frac{d}{dx}\left(c\frac{v}{T}\right) \qquad (1.17)$$

(recall that the pressure in the gas mixture is uniform). From (1.17) we obtain the equilibrium concentration of the light gas in the heavy gas:

$$c(x)\sim\frac{T(x)}{v(x)}\sim\sqrt{T(x)},\ \text{or}\ c(x)=c_0\sqrt{T(x)/T_0}. \qquad (1.18)$$

As is evident from (1.18), *in diffusion equilibrium the light gas tends to be concentrated in regions with high temperature.*

We turn now to the case when a heavy gas diffuses into a light one. Here, we assume that the concentration of the heavy gas is small in comparison with the concentration of the light gas. As before we assume that the mass of a heavy-gas molecule is much larger than the mass of a light molecule: $M_2 \gg M_1$.

Because the concentration of the heavy gas is small we can neglect collisions of heavy molecules with one another and assume that diffusion occurs solely due to collisions of heavy molecules with light molecules. The effect of the light molecules on the heavy ones can be treated as a frictional force **F**, analogous to the frictional force on a macroscopic body moving in a gas. In our case the force **F** arises because the concentration gradient dc/dx of the heavy gas induces a small velocity component **V** along the direction of the gradient. Obviously, **V** is small in comparison with the random thermal velocities of the heavy molecules and is small in comparison with the thermal velocities of the light molecules, which will be denoted by v, as before. Therefore we have $V \ll v$.

As noted above, the motion of a heavy molecule in a medium of light molecules is analogous to the motion of a macroscopic body in a viscous gas or liquid. Since the velocity of motion V is small, the relation between the frictional force and the velocity is linear

$$\mathbf{V}=\mu\mathbf{F}. \tag{1.19}$$

The quantity μ is called the *mobility*. We estimate this quantity.

According to Newton's second law, the force $\mathbf{F}$ is the total momentum per unit time transferred to a heavy particle by light particles colliding with it. In these collisions the velocity of the heavy molecule is practically unchanged because of its large mass and the nonrandom part of its velocity remains equal to $\mathbf{V}$. In each collision between a light molecule and a heavy one, the momentum transferred to the heavy molecule is of the order of the momentum of the light molecule $p = M_1 v$. More precisely, if the heavy molecule moves with velocity V, then in a head-on collision with a light molecule a momentum of order $M_1(v + V)$ is transferred to the heavy molecule and in a collision where the light molecule overtakes the heavy one the transferred momentum is of order $M_1(v - V)$. The total momentum from two such collisions is of order $M_1 V$ (if the motion of the heavy molecule is purely random it is equal to zero because of the random distribution of momenta of the light molecules colliding with the heavy molecule).

The number of collisions of a heavy molecule with light molecules per unit distance is $1/l$, where l is the mean free path of a heavy molecule. The collision frequency is equal to v/l, where v is the relative velocity between the light and heavy molecules in a collision and is practically equal to the velocity of the light molecule because of its small mass. Multiplying the momentum $M_1 V$ from two collisions by the collision frequency v/l, we obtain the momentum $M_1 Vv/l$ transferred by light molecules to a heavy one per unit time. This is also the required frictional force F. Hence

$$F \sim \frac{p}{l} V. \tag{1.20}$$

From (1.19) and (1.20) we obtain an estimate for the mobility:

$$\mu \sim \frac{l}{p}. \tag{1.21}$$

Substituting (1.1) for the mean free path l into (1.21), we find

$$\mu \sim (M_1 v n \sigma)^{-1} \sim (\rho v \sigma)^{-1}, \tag{1.22}$$

where $\rho = M_1 n$ is the *mass density* of the light gas. It can be taken as the total mass density of the gas because of the small concentration of the heavy gas. Substituting (1.3) for the average thermal velocity of a light molecule into (1.22), we obtain

$$\mu \sim \frac{1}{n\sigma\sqrt{M_1 T}}. \tag{1.23}$$

This equation can also be written in terms of the total pressure P of the gas mixture, which is determined by the ideal-gas equation for a rarefied gas $P=nT$ (the pressure in this case is created by the light gas):

$$\mu \sim \frac{1}{P\sigma}\sqrt{\frac{T}{M_1}}. \tag{1.24}$$

We determine now the diffusion coefficient of the heavy gas diffusing into the light gas. We first find a relation between the quantities μ and D. We write the diffusion flux $i = n_2 V$ of heavy molecules using (1.19) on the one hand and using the definition (1.6) on the other:

$$i=n_2V=n_2\mu F=-D\frac{dn_2}{dx} \tag{1.25}$$

(n_2 is the concentration of molecules of the heavy gas).

The force F and the concentration gradient dn_2/dx producing it can be related using the distribution of heavy molecules in a potential field $U(x)$ given by $F = dU/dx$. This distribution is well known from statistical physics (**the Boltzmann distribution**)

$$n_2(x) \sim \exp[-U(x)/T]. \tag{1.26}$$

Recall that the Boltzmann distribution is valid for a rarefied gas in an external field. Differentiating (1.26) with respect to x and dividing the resulting expression by (1.26), we find

$$\frac{1}{n_2}\frac{dn_2}{dx}=-\frac{1}{T}\frac{dU}{dx}=-\frac{F}{T}. \tag{1.27}$$

From (1.15) and (1.27) we obtain the required relation between D and μ:

$$D=\mu T. \tag{1.28}$$

This relation is called the **Einstein relation.**

Substituting (1.24) into (1.28), we obtain an estimate for the diffusion coefficient of a heavy gas diffusing into a light gas:

$$D \sim \frac{1}{P\sigma}\sqrt{\frac{T^3}{M_1}}. \tag{1.29}$$

Comparing (1.9) with (1.29), we see that the estimates for the diffusion coefficients of a heavy gas diffusing into a light gas and a light gas diffusing into a heavy gas are identical, in spite of the fact that the mechanisms of the diffusion process are completely different.

The relation (1.19) can also be used in the case of diffusion of a small concentration of light gas into a heavy gas. The quantity μ will then represent the mobility of a light molecule; V in this case will be the nonrandom part of the velocity of a light molecule directed along the concentration gradient (on the background of its much larger random thermal velocity v). Then (1.9) can be derived in the same way as (1.29). Indeed, the quantity M_1V is again the typical momentum transferred per collision, since the transferred momentum M_1v vanishes when averaging over the angles of the velocity vector $\mathbf{v}$ associated with the thermal motion of the light gas. The number of collisions per unit time is given by the same formula, $n\sigma v$, as in the case of diffusion of a heavy gas, only in this case $n \approx n_2$ (and not n_1, as in the case of diffusion of a heavy gas). Applying the Einstein relation (1.28) to the light molecules, we obtain (1.29). Therefore the equality of the estimates for the diffusion coefficients of a heavy gas diffusing into a light one and a light gas diffusing into a heavy one becomes obvious. In addition, we conclude that (1.29) is also valid when the mass of a molecule of the diffusing gas is comparable to the mass of a molecule of the solvent gas. In this case M_1 is either of the two masses. When these masses are identical D is called the *self-diffusion coefficient*. The phenomenon of *self-diffusion* can be observed if the diffusing molecules can be distinguished in some way from the solvent molecules (an example is the diffusion of excited molecules into a gas of molecules in the ground state).

Up to now we have assumed that the concentration of the diffusing gas is small. If this is not the case, then from the fact that the pressure P is uniform, i.e., does not depend on x, we have from the ideal-gas equation $P = (n_1 + n_2)T$

$$\frac{dn_1}{dx} = -\frac{dn_2}{dx}, \tag{1.30}$$

and hence the molecules of gases 1 and 2 diffuse in opposite directions. Since the estimates for the two diffusion coefficients are identical, it follows that the diffusion fluxes of gases 1 and 2 are comparable to one another.

When the masses M_1 and M_2 are the same order of magnitude, the expression (1.7) for the diffusion coefficient can be obtained simply from dimensional analysis.

Problem 1. Show that the drift velocity of oxygen molecules toward the Earth's surface under the force of gravity is of order

$$V \sim \frac{g}{n\sigma}\sqrt{\frac{M}{T}},$$

where g is the acceleration of free fall.

Problem 2. Show that the mean-square displacement of a given gas molecule due to multiple collisions with other gas molecules is given by the following function of the time t:

$$\langle r^2 \rangle \sim Dt.$$

Problem 3. In a gas there is a small admixture of molecules which can absorb laser radiation. The gas mixture flows in a channel with transverse dimension L. It is exposed to radiation in the transverse direction. As a result, after the gas passes through the irradiated zone a nonuniform distribution of the impurity is created. Show that to obtain a perceptible separation factor, it is necessary to have a radiation energy per molecule of order

$$\varepsilon \sim \frac{L}{l}\frac{c}{v}T.$$

Here, l is the mean free path of an impurity molecule in the gas, v is the thermal velocity of a molecule, and c is the speed of light.

Problem 4. Show that the diffusion coefficient of molecules in air under normal conditions is of order 10^{-5} m^2/s.

Problem 5. Show that if the repulsive force between molecules falls off slowly with distance then thermal diffusion proceeds such that the hotter regions are enriched in the heavier component of the gas.

1.3. Thermal conductivity of a gas

Our next example of a slightly nonuniform gas involves the presence of a small temperature gradient dT/dx in a gas of molecules of a single kind. This gradient disturbs the gas from the state of thermodynamic equilibrium and produces an energy flux q, which in hydrodynamics is related to the temperature gradient by the phenomenological relation (**Fourier's law**):

$$q=-\lambda\frac{dT}{dx}. \tag{1.31}$$

The quantity λ is called the thermal conductivity. The fact that (1.31) involves only the first derivative with respect to x and only to the first power is a consequence of the assumption that the gas is slightly nonuniform. In this case the neglected terms are of higher order in the small parameter l/L, where l is the mean free path and L is a characteristic length over which the temperature T varies.

The problem is to estimate λ. To do this we note that the energy flux q_+ is related to the flux i of gas particles by the equation

$$q_+=E(T)i=E(T)nv, \tag{1.32}$$

where $E(T)$ is the thermal energy per molecule. We can repeat the reasoning used to obtain (1.5) for the diffusion flux. The net energy flux from motion of molecules from left to right and from right to left through a unit area perpendicular to the direction of the temperature gradient is of order (in the absence of a net flux of particles):

$$q\sim nvl\frac{dE}{dx}\sim Cvl\frac{dT}{dx}. \tag{1.33}$$

Here, $C=ndE/dT$ is the heat capacity of the gas per unit volume.

From (1.33) and (1.31) we obtain an estimate for the thermal conductivity

$$\lambda\sim Cvl\sim nvl\sim\frac{1}{\sigma}\sqrt{\frac{T}{M}}. \tag{1.34}$$

Here, we have used the estimate (1.3) for the molecular velocity v. In addition, in obtaining (1.34) we used (1.1) for the mean free path and also used the fact that the heat capacity per molecule of a gas is of order unity.

As noted above, neutral molecules behave qualitatively as hard elastic spheres interacting with one another only in direct collisions (assuming the temperature is not too low; see Problem 2). The collision cross section is $\sigma\sim\pi d^2$, where d is a molecular diameter. Setting $d \sim 2 \times 10^{-10}$ m, we obtain $\sigma \sim 10^{-19}$ m^2 for the typical gas-kinetic cross section. In this case σ does not depend on the velocities of the colliding molecules and hence is also independent of the temperature of the gas T. Therefore we find from (1.34) that $\lambda \sim \sqrt{T}$. In addition, it is evident from (1.34) that at a given temperature T *the thermal conductivity does not depend on the pressure P of the gas or on its concentration n.*

It is easy to show that the estimate (1.34) remains qualitatively unchanged if we take into account rotational motion of the molecules in addition to their translational motion (the same statement is true for the estimates of the mobility and diffusion coefficient of the preceding section). Indeed, because the rotational motion is classical, the contribution to the heat capacity per molecule is of the same order as the contribution from translational motion (of order unity). This assertion is intimately connected with the well-known equipartition theorem of classical thermodynamics.

A temperature gradient creates a dissipative process in the gas: heat flows into a given volume of gas because of heat conduction. If the temperature gradient is not maintained by outside sources, the gas will transform into a state of thermodynamic equilibrium, i.e., uniform temperature. The characteristic time of this process τ_T over a length L is of order L/V, where V is the component of the molecular velocity along x due to the temperature gradient. From the derivation given above, it is evident that the estimate (1.12) is valid for V, as in the case of diffusion. Then we obtain the estimate $(L/l)^2(l/v)$ for the time τ_T. We see that τ_T is large in comparison with the mean free time $\tau = l/v$. Note that the mean free time characterizes energy relaxation, since in each collision the energy of a molecule changes by a quantity of the order of the energy itself.

Problem 1. Show that the upward heat flux q in the Earth's atmosphere due to the decrease of temperature with distance from the Earth's surface is of order 10^{-4} W/m^2. In estimating the temperature gradient of the atmosphere assume that the process of heat transport is very weak and hence the adiabatic relation between the temperature gradient and the barometric density gradient holds for the atmosphere.

Problem 2. At low temperatures the typical velocity of a molecule is small and long-range interactions between molecules are significant. For neutral molecules the long-range interaction falls off with the distance r between the molecules as r^{-6} (the van der Waals interaction). Show that the collision cross section of the molecules σ depends on the relative velocity v of the colliding molecules as $v^{-2/3}$ and that the thermal conductivity in this case depends on temperature as $T^{5/6}$.

Problem 3. For a gas of atoms with a small admixture of n_2 diatomic molecules per unit volume, estimate the ratio of the thermal conductivity due to heat transport by the atoms λ to that due to dissociation of the diatomic molecules λ_2. Assume that the dissociation energy E_D is large compared to the temperature T of the gas. The equilibrium number of molecules can be estimated using the Boltzmann formula as $n_2 \sim \exp(-E_D/T)$. Show that the ratio is of order

$$\frac{\lambda_2}{\lambda} \sim \frac{n_2}{n}\left(\frac{E_D}{T}\right)^2$$

(n is the concentration of atoms) and can be of order unity even if $n_2 \ll n$ because $E_D \gg T$. Use the fact that each atom in the heat flux carries an energy of order T, while each molecule carries an energy of order E_D.

Problem 4. Show that the independence of the thermal conductivity (1.34) on the gas pressure is a consequence of the approximation of binary collisions between molecules.

1.4. Viscosity of a gas

The viscosity of a slightly nonuniform gas is calculated in the same way as the thermal conductivity: the difference is that here the deviation from equilibrium is caused by a nonuniformity in the macroscopic velocity V of motion of the gas, rather than by a temperature gradient. As before, we assume that the nonuniformity is weak.

Let the velocity V be along the z axis and suppose it varies in the transverse direction x. Then momentum transfer arises between regions with different gas velocities. This transfer leads to frictional forces which retard the motion of the gas in the region with the higher velocity and tend to smooth out the velocity to an average value. Hence we have relaxation of a nonequilibrium state of the gas. In hydrodynamics the force F acting on a square of unit area in the yz plane perpendicular to the x axis, and directed opposite to the motion of the gas, has the form

$$F = -\eta \frac{dV}{dx}. \tag{1.35}$$

The quantity η is called the *dynamical viscosity.*

The frictional force F per unit area is in essence a pressure. It can also be considered as the viscous part of the momentum flux of the gas. Indeed, the momentum flux Π is the product of the momentum p per molecule by the flux i of these molecules, i.e., by the number of molecules passing through a unit area in unit time. The change of momentum per unit time is a force, and when divided by the area it represents a pressure, which is denoted above by F.

Note the analogy between the processes of diffusion, heat conduction, and viscosity. They are all the result of the same molecular mechanism, i.e., the direct transfer of molecules by means of molecular collisions and induced by a gradient in the system. The phenomenon of heat conduction can be considered as "diffusion" of thermal energy, whereas viscosity is diffusion of momentum.

We estimate the dynamical viscosity using the same method as was used above to estimate the thermal conductivity. The momentum flux Π is related to the flux of molecules i by a relation analogous to (1.32):

$$\Pi = M[v + V(x)]i. \tag{1.36}$$

This expression takes into account the motion of the gas as a whole with the macroscopic velocity $V(x)$. In the net momentum flux through a unit area perpendicular to the x axis, the first term in (1.36) drops out because it does not depend on x. The second term leads to the net momentum flux, which is calculated in analogy with (1.33):

$$F=\delta\Pi\sim Mil\frac{dV}{dx}\sim Mnvl\frac{dV}{dx}. \tag{1.37}$$

Comparing (1.35) with (1.37), we obtain an estimate for the kinematic viscosity:

$$\eta\sim Mnvl\sim\rho lv\sim\frac{1}{\sigma}\sqrt{MT}. \tag{1.38}$$

Here, we have used (1.3) for the average velocity of a molecule v. We see that the *dynamical viscosity does not depend on the gas pressure and increases with increasing temperature as* $T^{1/2}$.

In each collision the momentum of a molecule changes by a quantity of the order of the momentum itself, which leads to a randomization of the momentum in direction and magnitude in the process of approach to thermal equilibrium. Hence *the mean free time (1.4) is the momentum relaxation time.* The velocity is smoothed out over a macroscopic length L in a time which is larger than the mean free time by the factor $(L/l)^2$.

Problem 1. Show that for normal conditions the dynamical viscosity of air is of order 10^{-5} Pa s.

Problem 2. Obtain qualitatively the dependence of the dynamical viscosity on temperature in the form (*the Sutherland formula*)

$$\eta\sim(1+C/T)^{-1}/\sqrt{T},$$

where C is a constant (*the Sutherland constant*), starting from the temperature dependence of the collision cross section of neutral molecules.

1.5. Momentum and energy diffusion

In many kinetic processes the average change of the physical quantities in a single collision are small in comparison with the typical values of the quantities themselves. Such processes can be called *slow processes.* The time characterizing a slow process is much larger than the time between successive molecular collisions (the mean free time).

We have already considered an example of a slow process when we discussed the diffusion of a small amount of heavy gas into a light gas in Sec. 1.2. Because of the small concentration of molecules of the heavy gas we could neglect collisions of heavy molecules with one another. In a collision between a heavy molecule and a light one the momentum of the heavy molecule changes only slightly because of its large mass. Over a large number of collisions the nature of the change of momentum is like a diffusion of momentum. Indeed, in each collision the momentum of a given heavy molecule increases or decreases by a small amount. The mean-square momentum varies linearly with time, just like the mean-square coordinate of a particle in Brownian motion (see Problem 2 of Sec. 1.2).

We estimate the change of momentum for the example of the diffusion of a small amount of heavy gas into a light gas (see Sec. 1.2). Because the temperatures of the light and heavy components of the gas mixture are equal, the velocities of the heavy molecules are much smaller than the velocities of the light molecules. Therefore we can assume a heavy molecule is initially at rest in calculating the momentum transfer. According to the law of conservation of momentum, the change of momentum of the heavy molecule Δp_2 in a collision with a light molecule is equal to the change of momentum of the light molecule Δp_1. In the process of one such collision the direction of the momentum vector of the light molecule changes by an arbitrarily large angle, hence $\Delta p_1 \sim p_1$. Therefore in a single collision we have $\Delta p_2 \sim p_1$. The change Δp_2 can be positive or negative. The quantity $(\Delta p_2)^2 \sim p_1^2$ is always positive. Multiplying it by $n\sigma$, where n is the concentration of the light gas (which is practically equal to the concentration of all molecules of the gas mixture, since the concentration of the heavy component is small), and σ is the cross section for collisions of light molecules with heavy ones, we obtain the change in the square of the momentum of the heavy molecule per unit path length. Multiplying this expression by the velocity v_1 of a light molecule (in our case the relative velocity of the light and heavy molecules reduces to the velocity v_1 of the light molecule because of the small velocity of the heavy molecule), we obtain the change in the square of the momentum of the heavy molecule per unit time. Therefore the mean-square change in the momentum of a heavy molecule after time t is approximately

$$\langle (\Delta p_2)^2 \rangle \sim p_1^2 v_1 n\sigma t. \tag{1.39}$$

We consider first the change in the motion of the heavy molecule associated with its change of direction in space. Let θ be the deflection angle of the heavy molecule in a single collision. Then $\Delta p_2 = p_2\theta$. Using (1.39), we obtain the following estimate for the mean-square deflection angle after time t:

$$\langle \theta^2 \rangle \sim (p_1/p_2)^2 v_1 n\sigma t. \tag{1.40}$$

In a state of thermal equilibrium between the light and heavy components of the gas we have

$$p_1 \sim \sqrt{M_1 T}, \quad p_2 \sim \sqrt{M_2 T}, \tag{1.41}$$

where T is the temperature of the gas. Therefore we find from (1.40)

$$\langle \theta^2 \rangle \sim \frac{\sqrt{M_1 T}}{M_2} n\sigma t. \tag{1.42}$$

We see that the higher the temperature, the more rapid the diffusion of the deflection of the heavy molecule from its initial direction. Complete randomization of the direction of motion of the heavy molecule occurs after a time t, found from (1.42) by setting $\langle \theta^2 \rangle \sim 1$:

$$t \sim \frac{M_2}{\sqrt{M_1 T} n\sigma}. \tag{1.43}$$

We compare this time with the collision time (1.4), i.e., with the time between two successive collisions of a heavy molecule with light ones:

$$\tau \sim \frac{1}{n\sigma} \sqrt{\frac{M_1}{T}}. \tag{1.44}$$

Dividing (1.43) by (1.44), we obtain

$$\frac{t}{\tau} \sim \frac{M_2}{M_1} \gg 1. \tag{1.45}$$

Hence *complete randomization of the direction of motion of a heavy molecule occurs after M_2/M_1 collisions.* Recall that here M_1 is the mass of a light molecule and M_2 is the mass of a heavy one.

We considered the diffusion of the deflection angle of a heavy molecule. Now we turn to the diffusion of the magnitude of its momentum, which is determined by (1.39). Complete randomization of the magnitude of the momentum occurs when $\langle(\Delta p_2^2)\rangle \sim p_2^2$. Therefore we obtain the estimate (1.43) for the time of randomization of the magnitude of the momentum, and hence it is equal to the time of randomization of the deflection angle of the momentum of the heavy molecule.

We consider diffusion of the energy of the heavy molecule E_2. The energy ΔE_2 gained or lost in each collision of the heavy molecule is of order

$$\pm \Delta E_2 \sim \frac{p_2 \Delta p_2}{M_2} \sim \frac{p_2 p_1}{M_2} \sim \sqrt{\frac{M_1}{M_2}}\, T \ll E_2 \sim T. \tag{1.46}$$

In analogy with (1.39), after a time t we obtain

$$\langle(\Delta E_2)^2\rangle \sim (\Delta E_2)^2 v_1 n\sigma t \sim \frac{\sqrt{M_1 T^3}}{M_2}\, n\sigma t. \tag{1.47}$$

Complete randomization of the energy of the molecule occurs when the change in the energy becomes comparable to the energy itself (which is of order of the temperature T of the gas). It follows from (1.47) that the diffusion time in this case is again given by the estimate (1.43).

We consider now the diffusion of a small quantity of light gas into a heavy gas. The deflection angle of a light molecule in this case changes sharply in a single collision with a heavy molecule, and hence the time of randomization of the deflection angle is given by (1.44), where, however, n is now the concentration of all molecules (which is practically equal to the concentration of heavy molecules).

But because the collisions are elastic, the energy E_1 of the light molecule changes only slightly in a single collision. From conservation of energy, $\Delta E_1 = \Delta E_2$, where ΔE_2 is given by (1.46). Therefore we obtain the result (1.47) for $\langle(\Delta E_1)^2\rangle$ and the time of complete randomization of the energy is given by (1.43). It follows that the energy of the light molecule becomes completely randomized after $(M_2/M_1) \gg 1$ collisions [see (1.45)].

Problem 1. Show that the diffusion coefficient in momentum space for the momenta of heavy molecules in a gas of light molecules is of order

$$D_p \sim \sqrt{M_1 T^3}\, n\sigma,$$

where M_1 is the mass of a light molecule and n is the number of light molecules per unit volume.

Problem 2. An electron with energy E enters a gas of neutral molecules at temperature $T \ll E$ and is slowed down in the gas. Show that after $\sim M/m$ elastic collisions (m is the mass of the electron, M is the mass of a molecule) the energy of the electron drops to a quantity of order T, after which the temperatures of the electron and molecules are leveled by diffusion after $\sim M/m$ collisions.

Problem 3. Show that because the transverse momentum transfer vanishes when it is averaged over direction in a plane perpendicular to the motion of the molecule, the momentum transfer in molecular collisions is characterized not by the cross section σ, but by the so-called *cross section for momentum transfer*

$$\sigma_t = \int (1-\cos\theta)\, d\sigma,$$

where θ is the scattering angle. Show that for the case of hard spheres both cross sections are of the same order of magnitude.

1.6. Boundary between a gas and a wall

In gas dynamics the condition of thermal equilibrium between a gas and the container implies that the temperatures of the gas and container are equal. This condition assumes that the mean free path l is infinitesimal. In this section we consider the deviation from this condition due to the finite value of l.

We assume a temperature gradient in the gas in a direction perpendicular to the wall of the container. Let L be the characteristic distance over which the temperature varies. We will assume $L \gg l$, i.e., a slightly nonuniform gas. Then the deviation from equilibrium is small. The loss of thermal equilibrium leads to a deviation between the temperature of the gas and the temperature of the wall. Let this difference be δT. In analogy with (1.31), we can write

$$\delta T = g \frac{dT}{dx}. \tag{1.48}$$

The quantity g is called the *temperature jump coefficient*. From physical considerations it follows that the quantity g is of order l, since the gas molecules travel the final part of their way to the wall (a distance of order l) without collisions, and they therefore keep the temperature they had at a distance l from the wall.

From the point of view of macroscopic gas dynamics this effect can be taken into account phenomenologically by "moving" the wall back a distance g and applying the condition that the gas and wall temperatures be equal at the displaced position.

We see from (1.48) that the correction is first order in the small ratio l/L. It corrects the usual gas-dynamical solutions to order l/L.

The same reasoning can be used for the boundary condition on the velocity at the gas-wall boundary. In the hydrodynamics of a viscous gas or liquid this condition states that the velocity of a gas is equal to zero on the surface of the wall, i.e., a viscous gas "sticks" to the wall. When the mean free path l is nonzero there is a nonzero tangential velocity V_z on the wall surface. It results from a dependence of the tangential velocity $V_z(x)$ on the coordinate x normal to the wall. In analogy with (1.48) we can write

$$V_z = \xi \frac{dV_z(x)}{dx}. \tag{1.49}$$

The quantity ξ is called the *slip coefficient*. Like g, the quantity ξ is of order l.

From the point of view of macroscopic gas dynamics, this effect can be taken into account by "moving" the wall back a distance equal to ξ and imposing the condition that the macroscopic velocity of the gas must vanish at the displaced position.

As in the case of the temperature jump, the slip effect is first order in the parameter l/L for a slightly nonuniform gas. It is related to the viscosity of the gas.

A similar slipping effect occurs in the presence of a temperature gradient along the wall dT/dz (z is a coordinate running along the wall). In analogy with (1.49) we have

$$V_z = \mu_t \frac{dT}{dz}. \tag{1.50}$$

The quantity μ_t is called the *thermal slip coefficient*. It can be estimated from dimensional considerations using the fact that from (1.50) μ_t has the dimensions of velocity over temperature per unit length, and a quantity with these units must be constructed from M, v, and l. We find

$$\mu_t \sim l/(Mv). \tag{1.51}$$

Thermal slip can be generated by a nonuniformly heated wall, for example.

We see that *all of the conditions on the gas-wall boundary are first-order corrections in the parameter l/L to the solutions of the macroscopic gas-dynamical equations.* We estimate the analogous corrections arising inside the bulk of the gas because of higher-order terms in the expansion of the gas parameters in l/L. As an example, we consider the viscous part of the momentum flux $\delta\Pi$. To lowest order it is determined by (1.37) and is of order

$$\delta^{(1)}\Pi \sim \rho v l \frac{V}{L}. \tag{1.52}$$

Here, the index (1) means that we have used the first approximation in l/L.

Let $\delta^{(2)}\Pi$ be the viscous part of the momentum flux in the second approximation. It can be estimated by using the fact that in (1.52) v should be replaced by $v + V(x)$ in the second approximation, as was done in (1.36). Because

$$V(x) - V(0) \sim l \frac{dV}{dx} \tag{1.53}$$

and $V(0)$ can be set equal to zero with no loss of generality (physically, this corresponds to the statement that the kinetic characteristics of a gas do not depend on its macroscopic velocity, which has different values in different coordinate systems), we find from (1.52)

$$\delta^{(2)}\Pi \sim \rho l \frac{V}{L} l \frac{V}{L}. \tag{1.54}$$

Dividing (1.54) by (1.52), we obtain the relative correction:

$$\frac{\delta^{(2)}\Pi}{\delta^{(1)}\Pi} \sim \frac{l}{L}\frac{V}{v} \sim \left(\frac{l}{L}\right)^2. \tag{1.55}$$

Here, we have used (1.12) for the induced velocity V.

Hence the corrections to the equations of gas dynamics are quadratic in the parameter l/L and can therefore usually be omitted in comparison with the linear corrections in l/L resulting from the conditions on the gas-wall boundary.

Problem 1. Two containers of gas are at different temperatures (let the temperature difference be δT). The containers are joined by a long tube. Because of thermal slip, a pressure difference δP arises between the gases in the two containers (the so-called *thermomechanical effect*). Show that

$$\frac{\delta P}{P} \sim \frac{l^2}{R^2}\frac{\delta T}{T},$$

where R is the radius of the tube joining the containers.

Problem 2. Show that if a molecule colliding with a wall is absorbed by the wall and then emitted with the temperature of the wall (so-called *accommodation*), the temperature jump coefficient is equal to two times the mean free path.

Problem 3. Show that if the fraction c of molecules which completely lose the nonrandom component of their velocities when they are reflected by the wall is small ($c \ll 1$), then the slip coefficient ξ is of order l/c.

1.7. Highly rarefied gases

Up to now we have assumed that the mean free path is small in comparison to the characteristic length L of the nonuniformity in the gas. In this section we consider the opposite limit, when $l \gg L$. Obviously, this limit corresponds to a highly rarefied gas whose density is very small (a high vacuum). In this case the macroscopic equations of gas dynamics are in general inapplicable. However, kinetic theory remains valid, since its validity requires the much less restrictive condition $L \gg \langle r \rangle$, where $\langle r \rangle$ is the average distance between gas molecules (see Sec. 1.1). Then the gas can still be characterized by the average physical quantities and the relative fluctuations in these quantities are small. In particular, the gas can be characterized by the temperature T defined in terms of the average molecular velocity by (1.3).

We consider how to estimate the kinetic coefficients in the case $l \gg L$. We first consider a typical equilibrium process for a highly rarefied gas. We have two containers with gas joined by a tube whose diameter d is much smaller than the mean free path l. Let T_1, T_2 and P_1, P_2 be the temperatures and pressures in the two containers. The problem is to find a relation between these quantities in the equilibrium state.

When $l \gg d$ molecules leave one container and enter the other independently of one another, i.e., without colliding with one another. In collisions with the tube wall these molecules acquire the temperature of the wall if the number of such collisions is sufficiently large (i.e., we have the condition of accommodation; see Problem 2 of

Sec. 1.6). Therefore a molecule going into the other container will already have the temperature of that container and hence the temperatures T_1 and T_2 of the gases in the two containers must remain fixed. We determine a relation between the pressures P_1 and P_2 in this equilibrium state.

A number $\nu \sim nv$ of molecules collides with a unit area of the container wall per unit time. Here n is the concentration of molecules, and v is their average thermal velocity. Therefore the total number of molecules leaving one container and going into the other per unit time is

$$i \sim \nu d^2 \sim nvd^2. \tag{1.56}$$

A similar flux occurs in the opposite direction. In mechanical equilibrium these two fluxes must be equal. Therefore we obtain from (1.56) the equilibrium condition in the form

$$n_1 v_1 = n_2 v_2. \tag{1.57}$$

Using the ideal-gas equation $P = nT$ and (1.3) for the average velocity v, we obtain from (1.57)

$$\frac{P_1}{\sqrt{T_1}} = \frac{P_2}{\sqrt{T_2}}. \tag{1.58}$$

Hence in equilibrium we can have $P_1 \neq P_2$ (*the Knudsen effect*). As is clear from its derivation, the result (1.58) holds if the length of the tube connecting the two containers is large enough so that accommodation can occur.

We note that the ratio $K = l/L$ is called the *Knudsen number*. In this section we assume that $K \gg 1$.

We consider nonequilibrium processes when $K \gg 1$. Nonequilibrium processes are created by the boundary conditions (walls). Therefore, unlike the case $K \ll 1$, when $K \gg 1$ the kinetic coefficients are characteristics not only of the gas, but also depend on the conditions of the particular problem, i.e., on L. This will be shown in the examples discussed below. The main cause of this dependence is that the molecules collide mostly with the walls, rather than with each other.

As a first example we consider the thermal conductivity of a gas when $K \gg 1$. Assume that the gas is confined between two plates heated to different temperatures (T_1 and T_2). Let L be the distance between the plates; we assume that the plates are parallel to one another. In analogy with (1.31), we write the heat flux q from one plate to the other per unit area of surface as

$$q = \lambda \frac{T_2 - T_1}{L}. \tag{1.59}$$

The problem is to estimate the thermal conductivity λ assuming $l \gg L$. The difference between T_1 and T_2 is assumed to be much smaller than the values of T_1 and T_2 themselves. Then we can introduce an average thermal velocity v given by (1.3), assuming that the temperature T in this formula is an average of the close values T_1 and T_2. We will also assume the condition of accommodation at the wall, i.e., the molecules acquire the temperature of the wall when they collide with it.

The number of molecules incident upon a unit area of the wall is equal to the number nv of collisions with the wall. Each molecule carries a thermal energy of order T_1 (T_2) when it collides with wall 2 (wall 1). Therefore the net heat flux has the form

$$q \sim nv(T_2 - T_1). \tag{1.60}$$

Comparing (1.59) with (1.60), we find the thermal conductivity:

$$\lambda \sim nvL \sim PL/\sqrt{MT}. \tag{1.61}$$

Here P is the gas pressure and M is the mass of a molecule.

Comparing (1.61) with (1.34), which holds when $l \ll L$, we conclude that the transition from the case $l \ll L$ to the case $l \gg L$ corresponds to replacing l by L in (1.34), i.e., the distance between the plates takes the place of the mean free path. This is natural, since L plays the role of the mean free path when $l \gg L$. It follows from (1.61) that λ depends on L, i.e., on the conditions of the particular problem, and not only on the properties of the gas itself, as in the case $l \ll L$. This feature has already been mentioned above.

In addition, in contrast to (1.34), when $l \gg L$ the thermal conductivity at a given temperature depends on the gas pressure P. Comparing (1.34) with (1.61), we conclude that as the gas pressure P decreases, λ at first remains constant (for $l < L$) and then begins to decrease (when $l > L$).

An analogous method can be used to treat the following problem. Two parallel plates move parallel to one another with relative velocity V. It is required to find the force per unit area on the plates for a highly rarefied gas $l \gg L$.

As in the case of motion of a body in a dense gas, this force is a frictional force due to the viscosity of the gas. In analogy with (1.35),

$$F = \eta \frac{V}{L}. \tag{1.62}$$

The force acts in the direction opposite to the velocity V. The problem is to estimate the viscosity η when $l \gg L$. We use the solution (1.38) for the case $l \ll L$ and the result of the previous problem that when transforming from the case $l \ll L$ to $l \gg L$ it is necessary to replace l by L in the kinetic coefficients. Therefore we find

$$\eta \sim MnvL \sim PL\sqrt{\frac{M}{T}}. \tag{1.63}$$

Substituting this estimate into (1.62), we obtain the frictional force per unit area on the plate:

$$F \sim PV\sqrt{\frac{M}{T}} \sim \frac{V}{v} P \ll P. \tag{1.64}$$

We see that it is small in comparison with the static force on the walls due to gas pressure.

Finally, we estimate the quantity of gas passing through a cross section of pipe of diameter L per unit time due to temperature and pressure gradients assuming $l \gg L$. The mass of gas passing through a unit cross section of the pipe per unit time is $i = Mnv$, where M is the mass of a molecule, n is the concentration of molecules, and v is the average thermal velocity of a molecule, which is defined [see (1.3)] through the gas temperature T. According to the ideal-gas equation, $P = nT$, where P is the gas pressure. A mass flux i propagates along the pipe in both directions such that the net flux over a length dx along the pipe is

$$di = \frac{d}{dx}(MPv/T)dx. \tag{1.65}$$

What is the characteristic value of dx in (1.65)? In the case $l \ll L$ it would be the mean free path l [see (1.5)]. However, when $l \gg L$ the characteristic value of dx becomes of order L, since a molecule travels a distance of the order of the diameter L of the pipe between successive collisions with the inner wall of the pipe. Setting $dx \sim L$ and multiplying (1.65) by the cross-sectional area of the pipe L^2, we obtain the quantity of mass flowing through the pipe per unit time:

$$Q \sim L^2\, di \sim L^3 \frac{d}{dx}\left(P\sqrt{\frac{M}{T}}\right). \tag{1.66}$$

Assuming that the pressure and temperature vary uniformly along the length of the pipe from the values P_1, T_1 at one end to P_2, T_2 at the other, we can replace the derivative in (1.66) by the difference of the values of the temperature and pressure at the ends of the pipe divided by the length of the pipe a:

$$Q \sim \frac{\sqrt{M}L^3}{a}\left(\frac{P_2}{\sqrt{T_2}} - \frac{P_1}{\sqrt{T_1}}\right). \tag{1.67}$$

It follows from (1.58) and (1.67) that in a state of diffusion equilibrium we have $Q = 0$.

Problem 1. A plate moves parallel to its surface in a highly rarefied gas with velocity V. Show that it is heated to a temperature of order MV^2, where M is the mass of a gas molecule.

Problem 2. Two parallel plates separated by a distance $L \ll l$ move parallel to one another with a relative velocity V. The temperature T_1 of one plate is much larger than the temperature T_2 of the other. Show that a frictional force per unit area of order

$$F \sim nV\sqrt{MT_2}$$

acts on each of the plates. Assume complete accommodation.

Problem 3. Estimate the mass flow rate of a rarefied gas through a very small aperture of diameter $d \ll l$ in a thin wall. The pressure and temperature of the gas on different sides of the wall are P_1, T_1 and P_2, T_2, respectively. Show that the flow rate is of order

$$Q \sim \sqrt{M}d^2\left(\frac{P_2}{\sqrt{T_2}} - \frac{P_1}{\sqrt{T_1}}\right).$$

Problem 4. The result (1.67) for the mass flow rate of a rarefied gas through a long pipe assuming $l \gg L$ is valid when the molecules reflect diffusely from the walls, i.e., when they are absorbed and then emitted such that their nonrandom motion is completely lost. Show that if only a small fraction $c \ll 1$ of molecules are reflected diffusely, while the other molecules are reflected specularly, then the gas flow rate along the pipe is larger than (1.67) by a factor of order $1/c$.

Chapter 2

Weakly ionized gases

This chapter is concerned with the study of nonequilibrium processes in weakly ionized gases, in which the concentrations of free electrons and ions are small, and most of the gas is made up of neutral molecules. Hence only collisions of ions and electrons with neutral molecules need be taken into account, and collisions of electrons with one another and with ions and also collisions of ions with one another can be neglected.

As before, we will consider a rarefied (ideal) gas, where $n\sigma^{3/2} \ll 1$, where n is the concentration of gas molecules and σ is the scattering cross section of the particles (see the introduction to Chap. 1). In a weakly ionized gas the transport of charged particles is of interest, especially under the influence of an external electric field. Also, the formation and destruction of charged particles as a result of electron-molecule collisions are of interest.

2.1. Weakly ionized gas in an electric field

We consider a weakly ionized gas in a constant electric field E. The electric field is assumed to be weak in the sense that the average energy picked up by the electrons in the field is much smaller than the energy necessary to excite or ionize neutral gas molecules. Then, the collisions between electrons and molecules can be treated as elastic. We note that the electric current induced by the applied electric field is determined by the electrons; the ions can be neglected because of their large mass. Hence we will consider only the motion of electrons in an electric field E.

We suppose first that the electric field is weak in the sense that the energy picked up by the electron from the field over a time equal to the collision time is small in comparison with the energy lost or gained by the electron in a collision with a molecule. The first of these energies is of order eEl, where l is the mean free path of the electron, while the second, according to (1.46), is of order $\sqrt{m/M}T$, where m is the electron mass, M is the mass of a molecule, and T is the gas temperature (assumed to be the same for the electrons and neutral molecules). Hence we consider first processes satisfying the condition

$$eEl \ll \sqrt{\frac{m}{M}}\,T. \tag{2.1}$$

We estimate the electric current of electrons produced by the electric field. We have for the current density j

$$j = en_e V, \tag{2.2}$$

where n_e is the concentration of electrons, while V is their macroscopic (drift) velocity in the direction of E. According to (1.19), the quantity V is given in terms of the mobility μ of the electron as

$$V = e\mu E. \tag{2.3}$$

According to (1.21), the mobility μ is of order $l/\bar{p}$, where $\bar{p}$ is the momentum of the electron. In obtaining this estimate we used the result of Sec. 1.2 that the diffusion coefficient and hence the mobility [according to the Einstein relation (1.28)] have the same order of magnitude for the light and heavy components of a gas. In the problem considered here the "light" gas is the gas of electrons, while the "heavy" gas is the gas of neutral molecules, which makes up the major part of the medium in a weakly ionized gas.

Substituting (2.3) into (2.2) and defining the electrical conductivity σ_e by the usual relation $j = \sigma_e E$, we obtain

$$\sigma_e \sim n_e e^2 l/p. \tag{2.4}$$

Substituting (1.1) for the mean free path l and the estimate $\bar{p} \sim \sqrt{mT}$ for the thermal momentum of the electron into (2.4), we find

$$\sigma_e \sim \frac{ce^2}{\sigma\sqrt{mT}}. \tag{2.5}$$

Here, $c = n_e/n \ll 1$ is the concentration of electrons in the gas and n is the concentration of neutral molecules.

The ionic conductivity is negligibly small in comparison with (2.5) because the concentration of ions is the same order of magnitude as the concentration c of electrons, while the mass of the electron in the denominator of (2.5) must be replaced by the much larger mass of an ion M.

The result (2.5) is valid when the condition (2.1) is satisfied, which means that collisions are the dominant mechanism of energy exchange between particles; the theory of Chap. 1 was constructed on the basis of this mechanism, in particular, (1.21) for the mobility, which was used in the derivation of (2.5).

The fact that the conductivity σ_e in (2.5) is independent of E is a statement of Ohm's law. Hence *the condition (2.1) for a weak field E is the condition of applicability of Ohm's law for the current in a weakly ionized gas.*

As we have seen above, in the case of a weak field the electron acquires from a molecule or loses to it an energy of order $\sqrt{m/M}T$ in a single collision. The energy of the electron changes by a quantity of the order of the energy T itself after M/m collisions (see Sec. 1.5). Therefore the energy or momentum (magnitude) relaxation time of an electron is of order of the quantity (1.43), i.e.,

$$t \sim \frac{M}{m}\frac{l}{v} \sim \frac{M}{n\sigma\sqrt{mT}}, \tag{2.6}$$

which is large in comparison with relaxation time $\tau \sim l/v$ of the direction of the electron momentum.

We consider now the opposite case of a strong field in the sense of the inequality

$$eEl \gg \sqrt{\frac{m}{M}}\, T, \tag{2.7}$$

opposite to (2.1). First, we estimate the average energy of an electron ε in such a field. When (2.7) is satisfied we can assume (the proof is given below) that $\varepsilon \gg T$. The change in the energy of a molecule as a result of a collision with an electron is

$$\Delta\varepsilon = (\overline{\mathbf{p}} + \Delta\overline{\mathbf{p}})^2/(2M) - \overline{\mathbf{p}}^2/(2M).$$

Here, $\overline{\mathbf{p}}$ is the momentum of the molecule before the collision and $\Delta\overline{\mathbf{p}}$ is the change of momentum as a result of the collision between the molecule and an electron and is equal to the change of momentum of the electron. The quantity $\Delta\overline{\mathbf{p}}$ is of the order of the electron momentum itself $\overline{p}_e \sim (m\varepsilon)^{1/2}$, since the direction of motion of the electron changes by an arbitrary angle after the collision. For the same reason we have $\overline{\mathbf{p}} \cdot \Delta\overline{\mathbf{p}} = 0$ when averaged over many collisions. Therefore the average increase in the energy of the molecule after one collision $\Delta\varepsilon$ is of order $\overline{p_e^2}/M \sim m\varepsilon/M$. From conservation of energy, the same energy is lost by the electron per collision. Therefore the entire energy of the electron is lost after M/m collisions. Just as in the case of a weak field, the change in the energy of the electron reaches a quantity of the order of the energy itself after M/m collisions (see Sec. 1.5).

To calculate the average energy of an electron in a strong field we consider how the electron picks up energy in the field. After one free path it picks up an energy of order eEl. In the next free path this energy could be doubled if the electron continues to move in the direction of the field or could add up to zero if the electron goes in the opposite direction after a collision. Hence the accumulation of energy by the electron is a diffusion process. The mean-square energy picked up by the electron after M/m collisions is of order $(M/m)(eEl)^2$. This quantity should be set equal to the square of the average energy of the electron in the field, since in equilibrium the energy picked up from the field is compensated by the energy lost to collisions with moving molecules (this compensation is complete after about M/m collisions, as shown above). Hence the average energy of the electron is

$$\varepsilon \sim \sqrt{\frac{M}{m}}\, eEl. \tag{2.8}$$

It is evident from (2.7) and (2.8) that $\varepsilon \gg T$, which was already used above when we neglected the thermal energy of the molecule T. We note, however, that we assume that $\varepsilon \ll I$, where I is the ionization potential of the molecule. Then collisions between electrons and molecules can be treated as elastic and the gas remains weakly ionized.

We estimate now the nonrandom (drift) part of the velocity V of an electron due to the electric field. The average velocity v of an electron is obtained through the energy ε in the usual way:

$$v \sim \sqrt{\frac{\varepsilon}{m}} \sim (eEl)^{1/2}\left(\frac{M}{m^3}\right)^{1/4}. \tag{2.9}$$

When the energy ε changes by $\Delta\varepsilon$ the velocity changes by

$$\Delta v \sim \frac{\Delta \varepsilon}{\sqrt{m\varepsilon}}. \tag{2.10}$$

The change of energy $\Delta\varepsilon$ during one mean free path is of order eEl; therefore, the change of velocity in one mean free path is, according to (2.10), of order

$$\Delta v \sim \frac{eEl}{\sqrt{m\varepsilon}}. \tag{2.11}$$

Substituting (2.8) into (2.11), we find

$$\Delta v \sim \frac{\sqrt{eEl}}{\sqrt[4]{mM}}. \tag{2.12}$$

This expression is the drift velocity V of the electron picked up from the electric field E. Indeed, after a mean free path the electron suffers a collision, after which its direction of motion changes sharply and so the velocity V along E picked up in the next mean free path is again approximately given by (2.12). This is consistent with the statement of Sec. 1.5 that the relaxation time of the direction of velocity of a light particle is of the order of the mean free time $\tau \sim l/v$, i.e., the relaxation time is associated with a single mean free path. Hence we finally obtain

$$V \sim \Delta v \sim \frac{\sqrt{eEl}}{\sqrt[4]{mM}}. \tag{2.13}$$

This velocity is small in comparison with the average velocity v of the electron in a strong field given by (2.9). From (2.9) and (2.13), the ratio of the velocities is of order

$$\frac{V}{v} \sim \sqrt{\frac{m}{M}} \ll 1. \tag{2.14}$$

We now determine the conductivity of the electrons in a strong field. Substituting (2.13) into (2.2) and using the definition $j = \sigma_e E$, we obtain for the conductivity σ_e in a strong field:

$$\sigma_e \sim \frac{n_e e^2}{\sqrt[4]{mM}\sqrt{eEn\sigma}}. \tag{2.15}$$

We see that it depends on E, falling off with increasing E. Hence *Ohm's law does not hold for a strong field.*

Problem 1. Show that the drift velocity of the electrons in a weak field E is of order

$$V \sim \frac{eE}{n\sigma\sqrt{mT}}. \tag{2.16}$$

Problem 2. Assuming the polarization interaction $U = -\beta/r^4$ between an ion and a neutral molecule (interaction of a charge with a dipole), show that the ionic conductivity of a weakly ionized gas in a weak electric field is of order

$$\sigma_i \sim \frac{c_i e^2}{\sqrt{M\beta}}, \tag{2.17}$$

and does not depend on the gas temperature and is small in comparison with (2.5). Here, c_i is the concentration of ions in the gas and M is the mass of an ion.

Problem 3. Assuming that the mean free time τ of an electron in a weakly ionized gas is a constant, show that in a strong electric field E the average energy of the electron ε is of order

$$\varepsilon \sim M\left(\frac{eE\tau}{m}\right)^2, \tag{2.18}$$

and the conductivity is independent of E (Ohm's law) and is of order

$$\sigma_e \sim \frac{n_e e^2 \tau}{m}. \tag{2.19}$$

Problem 4. Space charge is introduced inside a certain volume of a weakly ionized gas. Show that it "diffuses away" after a typical relaxation time of $t \sim \varepsilon_0/\sigma$, where σ is the electrical conductivity and ε_0 is the permittivity of free space.

2.2. Collisional recombination

An equilibrium level of ionization in a weakly ionized gas is formed as a result of ionization of neutral molecules in collisions with each other and the inverse process of recombination of colliding charged particles (ions and electrons) accompanied by the formation of neutral molecules.

We consider the simplest system consisting of a single species of ion, in addition to neutral molecules and electrons. In equilibrium we have

$$\beta = \alpha n_e n_i. \tag{2.20}$$

Here, β is the number of electrons formed per unit volume and per unit time as a result of collisions of neutral molecules with one another. The right-hand side of (2.20) is the rate of loss of electrons due to their recombination with ions and is proportional to the concentrations of electrons n_e and ions n_i. The quantity α is called the *recombination coefficient*. In the case of singly charged ions we have $n_e = n_i$.

The process of recombination is slow in comparison with other processes in a weakly ionized gas because the formation of a neutral atom in a collision between an electron and an ion requires that the energy liberated in the collision be carried away by some means. In sufficiently dense gases this energy is transferred to a third particle (a neutral molecule) and we have a three-body collision. Recombination of this type is called *collisional recombination*.

We estimate the *collisional recombination coefficient*. An electron with an energy of the order of the temperature T of the gas moving in the field of a positive ion collides with a neutral molecule and its energy decreases. This process is an energy diffusion process, since in each collision the energy of the electron decreases by a very small fraction [the decrease is of order $\sqrt{m/M}T$; see (1.46)]. To recombine with the ion, the energy of the electron must decrease by an amount of the order of its initial energy T, i.e., the electron must collide with neutral molecules many times. These collisions will be binary collisions, since the electron is in a highly excited bound state in the field of the positive ion and moves together with the ion.

In order for the above discussion to be valid, the temperature T must be small in comparison with the ionization potential I of the molecule. Then the motion of the electron in its highly excited energy levels is quasiclassical (note that when $T \sim I$ the gas is highly ionized).

The result (2.20) holds in equilibrium. Suppose the system is disturbed from the equilibrium state by putting electrons into the gas with a concentration n_e much larger than the equilibrium concentration n_{0e}. Note that according to the Boltzmann distribution the equilibrium concentration of ions n_{0i} is equal to

$$n_{0i} \sim \exp(-I/T), \tag{2.21}$$

where I is the ionization potential of the molecule [here we do not consider the preexponential factor in (2.21), which can be significant]. For the condition of charge neutrality, the equilibrium concentration of electrons n_{0e} is also given by (2.21).

The decrease of n_e in time is described by

$$\frac{dn_e}{dt} = -\alpha n_e n_i. \tag{2.22}$$

Here, we have neglected the constant β [see (2.20)] because $n_e \gg n_{0e}$. In order of magnitude we have from (2.22)

$$n_e \sim (\alpha t)^{-1}. \tag{2.23}$$

To determine the recombination coefficient from (2.23) we need to estimate the characteristic values of the diffusion time t and electron concentration n_e. We consider t first.

This quantity is just the time required for the energy of an electron in highly excited levels to change by a quantity of the order of the temperature T. From (1.43) we have

$$t \sim \frac{M}{\sqrt{mT}n\sigma}. \tag{2.24}$$

Here, M is the mass of a molecule, m is the mass of the electron, n is the concentration of neutral molecules colliding with the electron, and σ is the cross section for collision of an electron with a neutral molecule.

We estimate now the typical value of n_e in (2.23). It is not the concentration n_{0e} of free electrons. The quantity n_e is rather the concentration of highly excited electrons bound to positive ions by Coulomb forces. According to the virial theorem the total energy ($\sim T$) of the electron in the Coulomb field is of the order of its potential energy $e^2/(\langle r\rangle\varepsilon_0)$. Here $\langle r\rangle$ is the average distance between the electron and the positive ion. Therefore

$$\langle r\rangle \sim e^2/(\varepsilon_0 T). \tag{2.25}$$

This distance should be the same order of magnitude as the distance between the given electron and neighboring ions and electrons and also between neighboring ions, since then the process of recombination requires only collisions with neutral molecules and does not require three-body encounters between electrons, ions, and neutral molecules: an ion is always near the electron.

Hence we obtain the following estimate for the concentration of electrons n_e (and ions) in (2.23):

$$n_e \sim \langle r \rangle^{-3} \sim T^3 \varepsilon_0^3 / e^6. \tag{2.26}$$

If the concentration of electrons is much smaller than this value, then in order for recombination to occur, an electron would have to approach an ion at the instant when a neutral molecule was incident on it; the probability of a three-body collision of this type is small. If the concentration of electrons is much larger than (2.26), a completely different recombination mechanism becomes important, in which electrons play the role of the third particle.

Substituting (2.24) and (2.26) into (2.23), we obtain an estimate for the recombination coefficient

$$\alpha \sim \frac{\sqrt{m}}{M} \frac{e^6 n \sigma}{\varepsilon_0^3 T^{5/2}}. \tag{2.27}$$

We see that α falls off with increasing temperature as $T^{-5/2}$. The condition for a weakly ionized gas is $n_e \ll n$, or from (2.26), $T \ll e^2 n^{1/3}/\varepsilon_0$. For example, for a gas under normal conditions this corresponds to a temperature less than 3500 K.

From (2.27) we see that α is proportional to n, and so collisional recombination is important in sufficiently dense gases.

In rarefied gases another recombination mechanism is important, namely *radiative recombination*, in which the liberated energy is carried off by a photon. Here, a third body (neutral molecule) is not needed for recombination. This process is also called *photorecombination*.

We estimate *the photorecombination coefficient* α_r. The decrease of the electron concentration n_e in time is described by (2.22); hence $\alpha_r n_i$ (n_i is the concentration of ions) is the frequency of collisions of the electrons with ions leading to recombination. This frequency is approximately $\sigma_r j$, where σ_r is the cross section for recombination accompanied by the spontaneous emission of a photon and j is the flux of ions through a unit cross section, i.e., $j = n_i v$. Here, v is the relative velocity of the electron and ion, which reduces to the velocity of the electron (1.3). Hence we obtain an estimate for the photorecombination coefficient

$$\alpha_r \sim v \sigma_r. \tag{2.28}$$

To calculate σ_r we use the *principle of detailed balance*:

$$\sigma_r \sim (k/\bar{p})^2 \sigma_i. \tag{2.29}$$

Here, σ_i is the cross section for ionization accompanied by the formation of a slow electron, $\bar{p}$ is the momentum of the electron, and k is the momentum of the emitted photon.

This principle holds if we assume that the probabilities of the forward and backward processes are equal. Let these probabilities be w_i and w_r, respectively.

Then $w_i = w_r$. The cross section of the process σ is obtained from the probability w (per unit time) by multiplying it by the number $d\nu$ of final states and dividing by the flux density j of incident particles.

In the case of ionization we have $d\nu_i \sim \bar{p}^2\,d\bar{p}$ and in the case of recombination $d\nu_r \sim k^2\,dk$. The flux density of electrons is equal to the velocity of the electron $v = \bar{p}/m$, whereas the flux density of photons is equal to the velocity of a photon c (i.e., the speed of light). Therefore

$$\frac{\sigma_r}{\sigma_i} \sim \frac{w_r\,d\nu_r}{v}\,\frac{c}{w_i\,d\nu_i} = \left(\frac{k}{p}\right)^2 \frac{mc\,dk}{p\,dp}\,.$$

From conservation of energy we have (I is the ionization potential) $\hbar\omega = I + \bar{p}^2/(2m) = ck$. Differentiating this relation, we find $mcdk = \bar{p}d\bar{p}$. Therefore we obtain for the ratio of the cross sections

$$\frac{\sigma_r}{\sigma_i} \sim \left(\frac{k}{p}\right)^2,$$

which gives (2.29). We used the fact that the number of ions is equal to the number of electrons and the number of photons.

Note that the photon frequency and momentum of the ionized electron are typically of the order of the characteristic atomic quantities. Then the ratio $k/\bar{p}$ is of order $e^2/(\varepsilon_0\hbar c) \sim 10^{-2}$, and therefore the cross section for photorecombination is four orders of magnitude smaller than the cross section for photoionization.

The cross section for ionization accompanied by the formation of a slow electron (recall that its energy T is small in comparison with the ionization potential I) in the case of dipole absorption of a photon is of order

$$\sigma_i \sim \frac{e^2}{\varepsilon_0\hbar c}\,a_B^2, \tag{2.30}$$

where a_B is the Bohr radius and $e^2/(\varepsilon_0\hbar c)$ is the *fine-structure constant.* Substituting (2.29) and (2.30) into (2.28), we obtain for the photorecombination coefficient

$$\alpha_r \sim \left(\frac{e^2}{\varepsilon_0\hbar c}\right)^3 \frac{a_B^3 I}{\hbar}\left(\frac{I}{T}\right)^{1/2}. \tag{2.31}$$

Problem 1. Consider binary recombination of a positive and negative ion in a weakly ionized gas by means of transfer of an electron from one ion to the other. Show that for low collision energies the cross section for recombination is of order de^2/ε, where ε is the energy of an ion and d is its radius. Show that the recombination coefficient for this process is of order

$$\alpha \sim de^2/\sqrt{MT}.$$

Problem 2. Consider three-body recombination of electrons and positive ions as a result of collisions between recombining electrons and free electrons in a gas. Show that if $n_e \ll \varepsilon_0^3 T^3/e^6$, the collisional recombination coefficient is of order

$$\alpha \sim \frac{e^{10} n_e}{\varepsilon_0^5 \sqrt{m} T^{9/2}} n_e.$$

2.3. Ambipolar diffusion

In the preceding sections of this chapter we have assumed that the concentration n_e of electrons is equal to the concentration n_i of ions (for singly charged ions) and hence the gas as a whole is electrically neutral. Suppose a difference in the concentrations of electrons and ions ($n_e - n_i \neq 0$) is created inside an element of volume of a weakly ionized gas (let it be a sphere of arbitrary radius R).

In practice, a concentration difference can arise if equal concentrations of electrons and ions are introduced in a region of radius R in a neutral gas (for simplicity we assume that the ions are singly charged). Because the electrons are much lighter than ions, they will diffuse much more rapidly than the ions and hence the condition of electrical neutrality will be violated. A radial electric field is produced in the region, which retards the dispersing electrons and therefore almost completely quenches the electronic current. On the other hand, the electric field accelerates the positive ions and therefore enhances diffusion of the ions. This enhancement of the diffusion of the ions is called *ambipolar diffusion.*

It is evident from the above discussion that ambipolar diffusion is diffusion of both electrons and ions. According to (1.6), the flux i_e of electrons created by a concentration gradient dn_e/dr is

$$i_e = -D_e \cdot dn_e/dr. \tag{2.32}$$

An electric field E creates flux of electrons $i'_e = n_e V$ in the opposite direction. Here, the drift velocity V, according to (1.19), is equal to $\mu_e eE$, where μ_e is the mobility of an electron. Setting $i_e = i'_e$, we obtain an expression for the electric-field strength:

$$E = -\frac{D_e}{e\mu_e n_e}\frac{dn_e}{dr} = -\frac{T_e}{en_e}\frac{dn_e}{dr}. \tag{2.33}$$

Here, we have used the Einstein relation (1.28). The quantity T_e is the *electron temperature*, which may be different in general from the *ion temperature* T_i (the time required to equalize the two temperatures is quite large; see the end of Sec. 3.7).

The flux of ions i is composed of a diffusion part and a part due to the electric field E:

$$i = -D_i \frac{dn_i}{dr} + n_i \mu_i eE. \tag{2.34}$$

Here, D_i and μ_i are the diffusion coefficient and mobility of the ions, and $n_i = n_e$ is the concentration of ions. From (1.28) and (2.33), we obtain from (2.34)

$$i = -\left(D_i + D_i \frac{T_e}{T_i}\right)\frac{dn_i}{dr}. \tag{2.35}$$

Therefore the diffusion coefficient D_a for ambipolar diffusion of ions has the form

$$D_a = \left(1+\frac{T_e}{T_i}\right) D_i. \tag{2.36}$$

In particular, when $T_i = T_e$ it is equal to twice the diffusion coefficient D_i. When $T_e \gg T_i$ diffusion of ions is greatly enhanced.

In a weakly ionized gas the concentrations of electrons and ions are small. However, electrons and ions are not distributed uniformly over the volume of the system. Each ion tries to pull one of the electrons toward it. As a result, the Coulomb field of the ion is screened. This does not mean that only one electron participates in screening the field of the ion. As we will see, the screening length is large in comparison with the average distance between ions and hence a large number of electrons and other ions participate in screening the field of the ion. We estimate this length (the so-called *Debye length* λ_{De}).

The energy of an ion is of the order of the temperature T of the gas. On the other hand, this energy is the sum of the energies of interaction of the given ion with all electrons and ions inside a sphere of radius λ_{De}. Here, we have used the fact that from the virial theorem for the Coulomb interaction, the total energy ($\sim T$) and Coulomb potential energy are of the same order of magnitude. The energy of interaction of an ion with an electron (or ion) is of order $e^2/(\lambda_{\mathrm{De}}\varepsilon_0)$. Multiplying this expression by the number of charged particles $n_e\lambda_{\mathrm{De}}^3$ in a sphere of radius λ_{De}, we obtain for the potential energy $n_e e^2\lambda_{\mathrm{De}}^2/\varepsilon_0$. Setting this estimate equal to the temperature T, we obtain an estimate for the Debye length:

$$\lambda_{\mathrm{De}} \sim \sqrt{\frac{T\varepsilon_0}{n_e e^2}}. \tag{2.37}$$

Suppose the condition of electrical neutrality is violated in a weakly ionized gas and there is a concentration difference $n_e - n_i \neq 0$. Then, because of diffusion of electrons this difference decreases rapidly and the condition of electrical neutrality is restored. We estimate the time for this process to occur. To restore electrical neutrality a given electron must leave the neighborhood of an ion and travel a distance of the order of a Debye length (2.37). According to (1.13), the time required for this process is of order

$$\tau_e \sim \lambda_{\mathrm{De}}^2/D_e, \tag{2.38}$$

where D_e is the diffusion coefficient of the electrons. From (1.8) it is of order

$$D_e \sim \frac{1}{n\sigma}\sqrt{\frac{T}{m}}, \tag{2.39}$$

where n is the concentration of neutral molecules in the gas and σ is the cross section for scattering of electrons by neutral molecules.

Substituting (2.37) and (2.39) into (2.38), we obtain

$$\tau_e \sim \frac{n\sigma\sqrt{mT}}{n_e e^2}. \tag{2.40}$$

Restoration of electrical neutrality is the fastest relaxation process in a weakly ionized gas.

Much slower is the process of ambipolar diffusion, in which the diffusion flux of electrons due to an electron concentration gradient cancels out the flux due to the difference between the electron and ion concentrations. This process requires a time of order

$$\tau_e' \sim \frac{R^2}{D_e}, \tag{2.41}$$

which is just the diffusion time for electrons due to a concentration gradient. Recall that here R is the characteristic length of the system over which the electron concentration is nonuniform. The diffusion formulas are only valid when $R \gg l$, where l is the mean free path of electrons (or ions; they are of the same order of magnitude). However, the condition $R \gg \lambda_{De}$ is not required; even if a disturbance is created in a small volume the concentration will be equalized rapidly because of rapid diffusion of electrons. We note that there is no restriction on the ratio of the Debye length λ_{De} and the mean free path of electrons l in a weakly ionized gas: the cases $\lambda_{De} > l$ and $\lambda_{De} < l$ can both occur. From the above derivation it is evident that the concept of the Debye length itself is valid when there are a large number of charged particles inside the Debye sphere, i.e., $n_e \lambda_{De}^3 \gg 1$ or

$$T \gg e^2 n_e^{1/3}/\varepsilon_0. \tag{2.42}$$

As we will see in Chap. 3, this is the condition for an ideal electron-ion subsystem. The condition (2.42) implies that the Debye length λ_{De} is much larger than the average distance between the charged particles and hence is even larger in comparison with the average distance between neutral particles.

Naturally, the process of ambipolar diffusion is the slowest process since its characteristic time is the diffusion time of the ions τ_i, which is roughly

$$\tau_i \sim R^2/D_i. \tag{2.43}$$

The diffusion coefficient of the ions D_i is, in analogy with (2.39):

$$D_i \sim \frac{1}{n\sigma} \sqrt{\frac{T}{M}} \ll D_e. \tag{2.44}$$

Problem 1. Show that if the electron and ion temperatures are different, then the Debye length is determined by the lowest of the two temperatures.

Problem 2. Show that the equilibrium difference of electron and ion concentrations leading to the electric field in the process of ambipolar diffusion is of order

$$|n_e - n_i| \sim T_e/(e^2 R^2).$$

Problem 3. Show that an external constant electric field penetrates a distance of the order of the Debye length into a weakly ionized gas.

Problem 4. Show that the decrease in ionization potential of a neutral atom (or molecule) in a weakly ionized gas is small in comparison with the thermal energy of the atom and is of order

$$\Delta I \sim e^2 n_e^{1/3}/\varepsilon_0.$$

Problem 5. Show that in the E layer of the Earth's atmosphere, where the electron concentration $n_e \sim 10^{11}\ \mathrm{m}^{-3}$ and the temperature $T \approx 300$ K, the Debye length is about 1 cm.

Chapter 3

Plasmas

The term plasma is assumed to mean a fully ionized gas. For simplicity, we assume a two-component plasma consisting only of electrons (charge $-e$) and singly charged positive ions (charge $+e$). Let n be the concentration of electrons (or ions).

We assume the plasma is rarefied (ideal). The condition for a rarefied plasma is that the average kinetic energy of a particle (ion or electron), which is of the order of the temperature T of the plasma, must be large in comparison with the average energy of interaction of the particle with its neighbor, which is $e^2/(\langle r\rangle\varepsilon_0)$. Here $\langle r\rangle$ is the average distance between particles in the plasma, i.e., $\langle r\rangle \sim n^{-1/3}$. Therefore we assume the condition

$$T \gg e^2 n^{1/3}/\varepsilon_0. \tag{3.1}$$

It places an upper bound on the density of the plasma.

The Debye screening length in a plasma is introduced in the same way as for the case of a weakly ionized gas [see (2.37)]:

$$\lambda_{\mathrm{De}} \sim \sqrt{\frac{T\varepsilon_0}{ne^2}}. \tag{3.2}$$

It determines the screening length of the Coulomb field of the particles in the plasma. Using (3.2), we can rewrite (3.1) in the form

$$\lambda_{\mathrm{De}} \gg \langle r\rangle \quad \text{or} \quad n\lambda_{\mathrm{De}}^3 \gg 1, \tag{3.3}$$

showing that the cloud of particles inside a Debye sphere contains a large number of particles.

In addition to the assumption of a rarefied plasma, we will also assume that the plasma is classical, i.e., obeys classical mechanics. This means that the energy of an electron must be large in comparison with the quantum energy of interaction of two electrons separated by a distance of order $\langle r\rangle$. According to the Heisenberg uncertainty principle, the typical momentum of an electron corresponding to this interaction is $p \sim \hbar/\langle r\rangle$, and the energy is of order

$$\varepsilon \sim p^2/m \sim \hbar^2/(m\langle r\rangle^2). \tag{3.4}$$

Because $\langle r\rangle \sim n^{-1/3}$, we obtain the inequality

$$T \gg \hbar^2 n^{2/3}/m \tag{3.5}$$

for the electrons to be treated as a classical system, rather than a quantum one. The mass m in (3.5) is of the electron; in the case of an ion of mass M, the inequality (3.5) is even more easily satisfied, since $M \gg m$. In other words, if the electron gas is classical, the ion gas will be too.

The ratio of the right-hand sides of (3.1) and (3.5) is of order

$$\frac{me^2}{\varepsilon_0 \hbar^2 n^{1/3}} \sim \frac{\langle r \rangle}{a_B}, \tag{3.6}$$

where $a_B = 4\pi\varepsilon_0\hbar^2/(me^2)$ is the Bohr radius and is equal to 0.05 nm. Because $\langle r \rangle \gg a_B$, we see that condition (3.1) is more restrictive than (3.5). Therefore a *rarefied (ideal) plasma is always classical.*

In kinetic theory equilibrium is approached by means of random collisions between particles. However, the mean free path of the particles in a plasma is not of the order of the Debye length λ_{De}. The screening of the cloud of particles is such that an ion is acted upon by the self-consistent mean field produced by the other ions and electrons. This mean field depends only on the position of the given ion, and must be treated as an external field, and not as the interaction potential of two colliding particles.

We first consider phenomena for which collisions between particles in the plasma are not important. In this case the mean free path of the particles in the plasma (it will be estimated later) is large in comparison with the dimensions over which the parameters disturbing the plasma from equilibrium change.

Since collisions are not taken into account, the evolution of the plasma in time does not approach statistical equilibrium. Hence the temperatures of the ionic and electronic components in the plasma may be completely different.

The motions of charged particles in a plasma, and especially the density of particles in different regions of space, are determined by the forces acting on these particles, i.e., electric or magnetic fields in the plasma. According to Maxwell's equations, the fields themselves are determined by the charge density distribution. Hence the *electric and magnetic fields in a plasma are self-consistent fields.* It is sometimes possible to treat the fields as given quantities (see Sec. 3.1, for example). However, this cannot be done when treating self-oscillations in a plasma, which produce oscillating electric fields (see Sec. 3.3).

We will consider only isotropic plasmas in which there is no preferred direction in space in the absence of external perturbations.

3.1. Plasma in an external electric field

In the presence of an applied electric field $\mathbf{E}$ a force $\mathbf{F} = e\mathbf{E}$ acts on the charged particles. However, here the problem is not simply an external field acting on a homogeneous plasma; the field penetrates into the plasma a distance no larger than a Debye length. We must take into account the electric fields produced by the charge density distribution of the plasma itself. We neglect the reverse effect of the motion of the particles on the fields.

The electric field $\mathbf{E}(\mathbf{r},t)$ can be expanded in a Fourier integral in the coordinates $\mathbf{r}$ and time t. Because of the linearity of the problem (in the approximation considered here) we can consider a single Fourier harmonic of this expansion and write

$$\mathbf{E}(\mathbf{r},t)=\mathbf{E}_0\exp[i(\mathbf{kr}-\omega t)]. \tag{3.7}$$

The quantity ω is the *frequency* of the field and $\mathbf{k}$ is its *wave vector*; the quantity $\lambda = k^{-1}$ is the *wavelength* of the field. We take the vector $\mathbf{k}$ along the direction of the x axis.

In most of this section we will consider longitudinal electric fields for which the field vector $\mathbf{E}$ is along the x axis, i.e., along the propagation vector $\mathbf{k}$ of the wave. In this case the field E can be written in the form

$$E(x,t)=E_0\exp[i(kx-\omega t)] \tag{3.8}$$

(at the end of this section we will consider a transverse electromagnetic field).

The field (3.8) leads to a polarization of the plasma, i.e., a dipole moment is created in the plasma. The dipole moment $\mathbf{P}$ per unit volume is called the *polarization*. In a weak field the polarization $\mathbf{P}$ is a linear function of the field: $\mathbf{P} = \varepsilon_0\chi(k,\omega)\mathbf{E}$. The quantity $\chi(k,\omega)$ is called the electric susceptibility. The dependence of χ on ω is called *frequency* (or time) *dispersion* while the dependence on k is called *spatial dispersion*.

For a weakly ionized gas considered in Chap. 2, spatial dispersion of the susceptibility can be neglected. Indeed, charged particles do not create polarization of the medium during a free path because they do not interact with it. Polarization arises only at the instant of collision with neutral molecules. If the wavelength is large in comparison with the molecular dimensions d ($kd \ll 1$), then one can practically always neglect the dependence of χ on k and therefore spatial dispersion does not occur.

However, in a plasma scattering by the self-consistent potential has a characteristic length of the order of the Debye length λ_{De}, which is much larger than d. Therefore spatial dispersion of the susceptibility can occur.

A similar analysis holds for frequency dispersion: it can be neglected in a weakly ionized gas if $\omega\tau_c \ll 1$, where τ_c is the collision time of a charged particle with a neutral molecule.

We first consider only the contribution of the electrons to the polarization of a plasma. Because the electronic and ionic components are independent, the polarization of the ions can be obtained by replacing the mass of the electron m by the mass of an ion M and the electron temperature T_e by the ion temperature T_i. The total polarization is the sum of the contributions from the electrons and the ions.

The unperturbed motion of an electron along the x axis is described by the equation $x = vt$, where $v = v_x$ is the unperturbed thermal velocity of the electron along the direction of the field $\mathbf{E}$. The field $\mathbf{E}$ perturbs the motion of the electron. Let the correction to the unperturbed coordinate be δx. From Newton's second law and (3.8), we have

$$m\frac{d^2\delta x}{dt^2}=-eE_0\exp[i(kv-\omega)t]. \tag{3.9}$$

On the right-hand side of (3.9) we have used the unperturbed coordinate of the electron, which is valid in the linear approximation in the weak field. The reverse effect of the perturbed motion on the field is higher order in the field and we neglect it.

Integrating (3.9), we find the displacement δx of the electron in the field:

$$\delta x = \frac{eE}{m(kv-\omega)^2},$$

or (3.10)

$$\delta \mathbf{r} = \frac{e\mathbf{E}}{m(\mathbf{kv}-\omega)^2}.$$

Multiplying the dipole moment of a single electron $-e\delta x$ by the number $n(v)dv$ of electrons per unit volume whose x component of the velocity lies in the interval $[v, v+dv]$, we obtain the polarization of the electronic component of the plasma:

$$d\mathbf{P} = -\frac{e^2 n(v)dv}{m(\mathbf{kv}-\omega)^2}\mathbf{E}. \tag{3.11}$$

Hence we obtain for the susceptibility of the electrons

$$\chi(k,\omega) = -\frac{e^2}{m\varepsilon_0}\int_{-\infty}^{\infty}\frac{n(v)dv}{(\mathbf{kv}-\omega)^2} = \frac{e^2}{mk\varepsilon_0}\int_{-\infty}^{\infty}\frac{n'(v)dv}{\omega-\mathbf{kv}}. \tag{3.12}$$

The last integral in (3.12) is obtained by means of an integration by parts. The symbol $f'(v)$ denotes the derivative df/dv.

The integrand of (3.2) has a pole at the point $\omega = \mathbf{k}\cdot\mathbf{v}$. We assume that the field was switched on at $t \to -\infty$. This can be taken into account in (3.8) by introducing the factor $\exp(\delta t)$ with $\delta = +0$. We see from (3.8) that this is equivalent to replacing the frequency ω by $\omega + i\delta$. Setting this substitution into (3.12), we find the following general expression for the electronic susceptibility:

$$\chi(k,\omega) = \frac{e^2}{mk\varepsilon_0}\int_{-\infty}^{\infty}\frac{n'(v)dv}{\omega-\mathbf{kv}+i\delta}. \tag{3.13}$$

We see that this quantity is complex. It determines the complex dielectric permittivity of the plasma $\varepsilon = 1 + \chi$. Equation (3.13) is written in SI (note the presence of the permittivity of free space ε_0.)

We consider approximations to the susceptibility (3.13) in different limiting cases. First, consider the case of a high-frequency field when $\omega \gg kv$. It is evident that kv can be neglected in comparison with ω in the denominator of the first integral on the right-hand side of (3.12). Using the normalization condition

$$\int n(v)dv = n \tag{3.14}$$

(n is the concentration of electrons), we find from (3.12)

$$\chi = -\frac{ne^2}{m\omega^2\varepsilon_0}. \tag{3.15}$$

We note that this result is also valid for transverse fields, since spatial dispersion is absent in this case.

The quantity

$$\omega_{pe}=\sqrt{\frac{ne^2}{m\varepsilon_0}} \tag{3.16}$$

is called the *plasma frequency* (the reason for this name will become clear below). Using (3.16), we can rewrite (3.15) in the form

$$\chi=-(\omega_{pe}/\omega)^2. \tag{3.17}$$

The quantity (3.16) is the electron plasma frequency. The ion plasma frequency can be defined in the same way; it is small in comparison with the electronic plasma frequency. In the limit $\omega \gg kv$ the contribution of ions to the susceptibility of the plasma will be smaller than the contribution of electrons by the factor M/m and therefore the dielectric permittivity is determined by the electronic component of the plasma.

It follows from (3.17) that the typical value of the frequency is $\omega \approx \omega_{pe}$. In this case the condition $\omega \gg kv$ implies that $\omega_{pe} \gg k(T_e/m)^{1/2}$ or $k\lambda_{De} \ll 1$, where λ_{De} is the Debye length. It follows from the discussion at the beginning of this section that when this condition is satisfied spatial dispersion of the susceptibility can be neglected. In other words, *spatial dispersion of the susceptibility can be neglected when the wavelength λ of the field becomes large in comparison with the Debye screening length*: $\lambda \gg \lambda_{De} \sim (\varepsilon_0 T_e/ne^2)^{1/2}$.

We see from (3.17) that frequency dispersion of the susceptibility χ disappears when $\omega\tau_e \gg 1$, where the time $\tau_e = \omega_{pe}^{-1}$ is the typical scattering time of an electron by the self-consistent potential (see also the discussion at the beginning of this section). In this case the plasma cannot follow the variation of the field in time.

We consider now the opposite limiting case of low frequency $\omega \ll kv$. Neglecting ω in the second integral on the right-hand side of (3.12), we obtain

$$\chi_e \sim \frac{n(v)e^2}{mk^2 v\varepsilon_0}. \tag{3.18}$$

From (3.14) we have $n(v) \sim n/v$. The quantity $v \sim (T_e/m)^{1/2}$ is the thermal velocity of an electron. Hence we finally obtain from (3.18)

$$\chi_e \sim \frac{ne^2}{k^2 T_e \varepsilon_0} \sim \frac{1}{(k\lambda_{De})^2}. \tag{3.19}$$

When $k\lambda_{De} \gg 1$ spatial dispersion of the susceptibility disappears because the plasma does not respond to such rapid variations of the field in space: its behavior is determined by the average zero-point field in space.

It follows from (3.19) that in the limit $\omega \ll kv$ frequency dispersion of the susceptibility disappears and we have the limit of the static susceptibility, independent of the frequency of the field.

When $\omega \ll kv$ the ionic part of the susceptibility can be taken into account by replacing the electron temperature T_e in (3.19) by the ion temperature T_i. Using $n_i = n_e = n$, we find

$$\chi_i \sim \frac{ne^2}{k^2 T_i \varepsilon_0} \tag{3.20}$$

for the ionic part of the susceptibility. Comparing (3.19) and (3.20), we conclude that, in contrast to the high-frequency limit, in the low-frequency limit $\omega \ll kv$ the contributions of the electronic and ionic components of the plasma to the susceptibility are comparable (assuming comparable temperatures $T_e \sim T_i$).

The results (3.19) and (3.20) were obtained for a longitudinal electric field. For electromagnetic waves the case $\omega \ll kv$ normally cannot occur because for a transverse electromagnetic field we have the dispersion relation

$$\omega \sim kc, \tag{3.21}$$

where c is the speed of light. For the case considered here of nonrelativistic electrons, we obviously have $c \gg v$ and therefore we always have $\omega \gg kv$.

Problem 1. Show that in the high-frequency limit $\omega \gg kv$ the correction to (3.15) for χ includes spatial dispersion and is of order

$$\Delta\chi \sim \frac{ne^2 kT_e}{m^2 \omega^4 \varepsilon_0}.$$

Problem 2. Show that in the low-frequency limit $\omega \ll kv$ the correction to (3.19) for χ includes frequency dispersion and has the form

$$\Delta\chi \sim \frac{ne^2 m\omega^2}{k^4 T_e^2 \varepsilon_0}.$$

Problem 3. Show that when the electronic component of the plasma is highly degenerate (i.e., $T_e = 0$) the electric susceptibility for $\hbar k \ll mv_F$ in the low-frequency limit $\omega \ll kv_F$ is of order

$$\chi \sim \frac{ne^2}{mk^2 v_F^2 \varepsilon_0}.$$

Here, v_F is the velocity of an electron on the Fermi surface and is related to the concentration n of plasma electrons by the equation

$$mv_F \sim \hbar n^{1/3}.$$

3.2. Landau damping

In this section we consider the imaginary part of the electric susceptibility of the plasma (3.13), which arises because of the pole at the point $v = \omega/k$:

$$\mathrm{Im}\,\chi = \frac{\pi e^2}{mk^2 \varepsilon_0} n'(v) \bigg|_{v=\omega/k}. \tag{3.22}$$

Estimating the derivative $n'(v)$ as

$$n'(v) \sim n(v)/v \sim n/v^2 \sim mn/T_e \tag{3.23}$$

for velocities of the order of the thermal velocity of the electron (1.3), and substituting this estimate into (3.22), we obtain

$$\mathrm{Im}\,\chi \sim \frac{ne^2}{k^2 T_e \varepsilon_0} \sim (k\lambda_{\mathrm{De}})^{-2}. \tag{3.24}$$

The estimate (3.23) is correct when the velocity ω/k is of the order of the thermal velocity $(T_e/m)^{1/2}$, i.e., $\omega \sim k(T_e/m)^{1/2}$.

If the frequency ω is higher than this value, then the quantity $f(v)$ falls off rapidly with increasing velocity, and hence the imaginary part of the susceptibility also falls off rapidly. For small frequencies ω the quantity $\mathrm{Im}\,\chi$ depends on the particular form of $n(v)$; however, $\mathrm{Im}\,\chi$ does not exceed the estimate (3.24). When (3.24) holds, the imaginary part of the electric susceptibility is of the same order of magnitude as the real part (3.19). For high frequencies $\omega \gg k(T_e/m)^{1/2}$ the quantity $n(\omega/k)$ falls off rapidly and the imaginary part of the suceptibility becomes small in comparison with the real part.

The estimate (3.24) is also correct for ions if we replace T_e by T_i and consider the frequency region $\omega \sim k(T_e/M)^{1/2}$, where M is the mass of an ion. This frequency region is much lower than the typical electron frequencies.

It is not difficult to show that in the case of a transverse electromagnetic field the imaginary part of the susceptibility vanishes because the ratio ω/k is equal to the phase velocity of light in the plasma c_p, which is larger than the speed of light in vacuum.

We show that physically the imaginary part of the susceptibility determines the energy dissipation of the electric field in the plasma. According to the Joule–Lenz law, the energy transformed into heat per unit time and per unit volume is

$$Q = jE = E\frac{dP}{dt}. \tag{3.25}$$

The last relation in (3.25) follows from the well-known fact that the displacement current density j is equal to the time derivative of the polarization P of the medium: $j = dP/dt$.

The real polarization P is related to the electric field E given by (3.8) by

$$P = \varepsilon_0(\chi E + \chi^* E^*). \tag{3.26}$$

The quantity E in (3.25) must also be real, and therefore should be replaced by $E + E^*$, where E is given by (3.8). Differentiating (3.26) with respect to time, we find

$$\frac{dP}{dt} = i\omega\varepsilon_0(-\chi E + \chi^* E^*). \tag{3.27}$$

Substituting (3.27) into (3.25) and averaging Q over time, we find

$$Q = i\omega\varepsilon_0(-\chi + \chi^*)|E|^2 = 2\omega\varepsilon_0\,\mathrm{Im}\,\chi|E|^2. \tag{3.28}$$

Hence the transformation of electrical energy into heat is determined by the imaginary part of the electric susceptibility $\mathrm{Im}\,\chi$.

It follows from (3.22) that energy dissipation is due to electrons whose velocity v in the direction of propagation of the field (the direction of the vector **E**) is equal to the phase velocity of the wave ω/k. We say that these particles move in phase with the wave. In a reference frame moving with the particle the electric field **E** is stationary and therefore the field does work on these particles which do not average out to zero over a cycle, as in the case of particles whose velocity is not equal to the phase velocity of the field; for these particles the field oscillates in a reference frame moving with the particle.

Since this mechanism of energy dissipation is not associated with particle collisions, it is fundamentally different from the usual energy absorption in processes such as heat conduction, viscosity, and diffusion, and it does not lead to an increase in the entropy. This mechanism of wave energy dissipation in a plasma is called *Landau damping.*

Problem 1. Show that when the electronic component of the plasma is highly degenerate (i.e., $T_e = 0$) and when $\hbar k \ll mv_F$ and $\omega < kv_F$, the imaginary part of the electric susceptibility is of order

$$\mathrm{Im}\,\chi \sim ne^2\omega/(mk^3v_F^3\varepsilon_0),$$

where v_F is the velocity of an electron on the Fermi surface and is related to the concentration of electrons by the equation $mv_F \sim \hbar n^{1/3}$.

3.3. Plasma waves

In the two preceding sections we considered the reaction of the plasma to an external electric field. When the field is turned off the reaction of the medium also disappears. However, if the frequency ω of the longitudinal electric field is such that the dielectric permittivity $\varepsilon(\omega) = 0$, then Maxwell's equations in the plasma are satisfied when the electric displacement $\mathbf{D} = 0$, while the electric field $\mathbf{E} \neq 0$. Indeed, the equation $\mathrm{curl}\,\mathbf{E} = 0$ is satisfied identically because the field $\mathbf{E} = E_x(x)\mathbf{i}_x$ is longitudinal, while the equation $\mathrm{div}\,\mathbf{D} = 0$ is satisfied because $\varepsilon(\omega) = 0$.

Hence waves with frequency ω for which $\varepsilon(\omega) = 0$ can propagate in a plasma. Returning to the case $\omega \gg kv$ it follows from (3.17) that $\varepsilon = 0$, i.e., $\chi = -1$ when $\omega = \omega_{pe}$. The plasma frequency ω_{pe} is defined by (3.16). In the opposite limit $\omega \ll kv$, it follows from (3.18) that the susceptibility χ is positive and therefore the dielectric permittivity ε cannot be equal to zero.

Only the real part of the susceptibility was taken into account in (3.17). When the imaginary part $\mathrm{Im}\,\chi$ is also taken into account, the condition $\varepsilon(\omega) = 0$ has the form

$$1 - \left(\frac{\omega_{pe}}{\omega}\right)^2 + i\,\mathrm{Im}\,\chi = 0. \tag{3.29}$$

The value of ω satisfying (3.43) is complex. Because the imaginary part $\mathrm{Im}\,\chi$ is small when $\omega \gg kv$ (see Sec. 3.2), we find, setting $\omega = \omega_{pe} + i\,\mathrm{Im}\,\omega$ into (3.29)

$$\mathrm{Im}\,\omega \sim \omega_{pe}\,\mathrm{Im}\,\chi. \tag{3.30}$$

The imaginary part Im ω is the damping constant of longitudinal electric waves varying as

$$E \sim \exp(-i\omega t) = \exp(-i\omega_{pe}t - \text{Im}\ \omega t). \quad (3.31)$$

The wavelength of the wave is infinite. The oscillations can be pictured as follows. We displace all of the electrons a small distance in the same direction. A macroscopic electric field then arises in the plasma along this direction, which tries to restore the electrons to their initial position. Oscillations of the electrons under such a field represent plasma oscillations. Because the direction of the oscillations is along the direction of the oscillating electric field, they are longitudinal oscillations.

Because of their large mass, the ions are not able to oscillate like the electrons, and remain fixed. According to Newton's second law, the displacement δx of the electrons is related to the amplitude of the oscillating electric field E by the equation

$$m\omega_{pe}^2 \delta x \sim E. \quad (3.32)$$

According to (3.22), damping of the plasma oscillations becomes stronger as the wavelength of the wave decreases (or the wave number k increases). In particular, when

$$k = \omega/v \sim \omega_{pe}\sqrt{m/T_e} \sim \lambda_{\text{De}}^{-1} \quad (3.33)$$

then it follows from (3.24) and (3.30) that Im $\chi \sim 1$ and Im $\omega \sim \omega_{pe}$ and so plasma oscillations disappear because the imaginary part of the frequency is of the same order of magnitude as the real part and the oscillations damp out after approximately one oscillation. This condition corresponds to a wavelength of the order of the Debye length.

We consider qualitatively how plasma oscillations interact with individual electrons of the plasma. Energy exchange between them can occur even in the absence of collisions. It is simplest to analyze the problem in a coordinate system in which the plasma wave is at rest. As in the case of the discussion at the end of the preceding section, we consider electrons whose velocity is close to the phase velocity of the wave. However, here we neglect Landau damping. In a coordinate system moving with the wave, the energy of such an electron is small and therefore it oscillates with a finite amplitude inside the potential well produced by the electric field of the wave, in spite of the small oscillation amplitude of the wave. Energy is exchanged at the instant when the electron reaches the sides of the potential well and changes direction. Electrons whose velocities differ significantly from the phase velocity of the plasma wave are not captured by the wave; exchange of energy by these electrons with the wave is very small in comparison with that of electrons whose velocities are close to the phase velocity. Hence in the discussion below we will not be interested in the electrons whose velocities differ significantly from the phase velocity of the wave.

The exchange of energy between a captured electron and the wave obviously averages out to zero after two collisions with the sides (left and right) of the potential well. However, if the captured electron collides with other electrons more often than it moves from one side of the potential well of the wave to the other, then there is a net energy exchange between the electron and the wave. It follows from physical considerations that the sign of this energy exchange is such that the electron picks up energy from the wave, i.e., plasma oscillations damp out with time because of the interaction with electrons. In essence, this mechanism is just Landau damping, when the distribution of electrons approaches an equilibrium Maxwellian distribution. We saw in Sec. 3.2 that Landau damping is due to electrons whose velocities are equal to the phase velocity of the wave.

The above discussion was based on classical mechanics. In quantum mechanics the plasma oscillations must be quantized. According to the quantization rules for the harmonic oscillator, the energy of a quantum corresponding to plasma oscillations (called a *plasmon*) is $E_p = \hbar\omega_{pe}$, where $\hbar$ is Planck's constant and the plasma frequency ω_{pe} is given by (3.16). Quantization of the motion changes the equilibrium Maxwellian distribution of electrons in the plasma. If the energy of an electron is equal to the energy of one or several plasmons, then the electron can excite plasma oscillations and the energy of the electron is transformed into the energy of a plasmon. The number of electrons with these energies becomes less than the number corresponding to an equilibrium Maxwellian distribution. Therefore there are dips in the equilibrium electron distribution function at multiples of the plasmon energy E_p. The typical width of a dip is of order $\hbar \, \mathrm{Im}\, \omega$, where $\mathrm{Im}\,\omega$ is given by (3.30).

In a weakly ionized gas damping of plasma oscillations is due to collisions between electrons and neutral molecules of the gas. In this case the quantity $\mathrm{Im}\,\omega$ in (3.31), which corresponds to damping of plasma oscillations, is of the order of the collision frequency $\nu \sim \tau^{-1}$ of an electron with neutral molecules. The quantity τ is the mean free time of an electron and is given by (1.4). Hence

$$\mathrm{Im}\,\omega \sim \nu \sim n\sigma(T/m)^{1/2}. \tag{3.34}$$

Here, n is the concentration of neutral molecules, T is the gas temperature, and σ is the cross section for collision of electrons with molecules. Plasma oscillations exist when $\omega_{pe} \gg \nu$.

As noted above, plasma waves exist when $\omega \gg kv$. Landau damping is due to electrons on the far tail of the equilibrium distribution, i.e., electrons with velocities

$$v = \omega/k = \omega_{pe}/k \gg \sqrt{T_e/m}.$$

On the other hand, the collisional mechanism of damping, resulting from collisions between electrons and the neutral gas molecules, is determined by electrons with the thermal velocity $v \sim (T_e/m)^{1/2}$.

The above results are for longitudinal plasma waves. We turn now to transverse plasma waves. In the case of a transverse field the equation $\mathrm{div}\,\mathbf{D} = 0$, where $\mathbf{D}$ is the electric displacement, is satisfied automatically because the vectors $\mathbf{D}$ and $\mathbf{k}$ are perpendicular and does not require the condition $\varepsilon(\omega) = 0$. The dispersion relation for transverse waves has the well-known form $\omega = kc_p$, where c_p is the speed of the wave (speed of light) in the medium and is related to the speed of light c in a vacuum by the equation $c_p^2 = c^2/\varepsilon$. Here, $\varepsilon < 1$ is the dielectric permittivity of the plasma. The velocity $c_p > c$ has already been mentioned above (see Sec. 3.2).

Therefore we have $\omega \gg kv$, since the electrons are nonrelativistic and therefore their velocities are small in comparison with the speed of light. Hence we conclude that Landau damping does not exist for transverse waves. Using (3.17), we obtain the following relation between ω and k for transverse electron plasma oscillations in the absence of spatial dispersion:

$$\omega^2=\omega_{pe}^2+k^2c^2 \tag{3.35}$$

instead of the relation $\omega = \omega_{pe}$ for longitudinal plasma waves. If $kc \gg \omega_{pe}$, then it follows from (3.35) that the effect of the medium becomes negligible.

Problem 1. Show that the relative k-dependent correction to the frequency of longitudinal plasma oscillations $\omega = \omega_{pe}$ is of order $(k\lambda_{De})^2$, where λ_{De} is the Debye screening length for the electrons, i.e., $\lambda_{De} = (\varepsilon_0 T_e/ne^2)^{1/2}$.

Problem 2. Show that a low-frequency electromagnetic wave $[\omega \ll k(T_e/m)^{1/2}]$ penetrates into the plasma a distance of order

$$\delta \sim \left(\frac{T_e}{m}\right)^{1/3} \left(\frac{c^2}{\omega\omega_{pe}}\right)^{1/6},$$

where ω_{pe} is the plasma oscillation frequency (3.16).

Problem 3. Show that the energy of a plasmon in a metal is of order 5–10 eV.

Problem 4. Show that a transverse electromagnetic wave whose frequency ω is slightly less than the plasma frequency ω_{pe}, but is the same order of magnitude, penetrates into the plasma a distance of the order of the wavelength $\lambda \sim c/\omega_{pe}$.

3.4. Ion-sound waves

Up to now we have considered the limiting cases $\omega \ll kv_e, kv_i$, or $\omega \gg kv_e, kv_i$, where v_e and v_i are the thermal velocities of the electrons and ions in the plasma, respectively. In this section we consider longitudinal electric oscillations whose frequencies ω and wave numbers k satisfy the inequality

$$v_i \ll \omega/k \ll v_e. \tag{3.36}$$

In addition, we assume $T_i \ll T_e$, which tends to reinforce the inequality (3.36), but which is required in the discussion below, independently of (3.36). Recall that v_e and v_i are given by the relations

$$v_e \sim \sqrt{T_e/m}, \quad v_i \sim \sqrt{T_i/M}. \tag{3.37}$$

Here, M is the mass of an ion. When (3.36) holds electron plasma oscillations do not occur. The contribution of the electrons to the electric susceptibility is calculated in the low-frequency limit (3.19), while the contribution of the ions is calculated in the high-frequency limit (3.17), where ω_{pe} is replaced by the ion plasma frequency

$$\omega_{pi}=\sqrt{\frac{ne^2}{M\varepsilon_0}}. \tag{3.38}$$

Using the fact that the contributions of the electronic and ionic components to the polarization of the medium are independent, we obtain for the electric susceptibility

$$\chi = -\left(\frac{\omega_{pi}}{\omega}\right)^2 + \left(\frac{C}{k\lambda_{\rm De}}\right)^2, \tag{3.39}$$

where C is of order unity. Setting χ equal to -1 (i.e., setting the dielectric permittivity $\varepsilon = 1 + \chi$ to zero), we obtain the dispersion relation for longitudinal electric waves:

$$\omega^2 = \frac{\omega_{pi}^2}{1 + (C/k\lambda_{\rm De})^2}. \tag{3.40}$$

Waves of this type are called *ion-sound waves.*

In the long-wavelength limit when $k\lambda_{\rm De} \ll 1$, we find from (3.40)

$$\omega \sim \omega_{pi}\lambda_{\rm De}k. \tag{3.41}$$

The dependence $\omega \sim k$ is the trademark of sound waves and the proportionality constant between ω and k is the speed of sound c_s

$$c_s \sim \omega_{pi}\lambda_{\rm De} \sim \sqrt{\frac{\rm T_e}{\rm M}} \ll \rm v_e. \tag{3.42}$$

We note that for a Maxwellian plasma $C = 1$ and hence the proportionality constant in (3.39)–(3.42) is equal to unity. We also note that the speed of ion sound is small in comparison with the thermal velocity of the electrons v_e. In the acoustic limit the condition (3.36) takes the form

$$\frac{T_i}{M} \ll \frac{T_e}{M} \ll \frac{T_e}{m}. \tag{3.43}$$

It is satisfied when $T_i \ll T_e$, which was assumed above.

If, on the other hand, the wavelength is small, such that $k\lambda_{\rm De} \gg 1$, then we obtain from (3.40) $\omega = \omega_{pi}$. These ion plasma waves are analogous to the longitudinal electron plasma waves considered in the preceding section. The condition (3.36) for the case of ion waves takes the form

$$v_i \ll (\omega_{pi}/k) \ll v_e \tag{3.44}$$

or

$$k\lambda_{\rm Di} \ll 1, \quad k\lambda_{\rm De} \gg \sqrt{\frac{m}{M}}, \tag{3.45}$$

where the Debye screening lengths for electrons and ions are defined as

$$\lambda_{\rm De} \sim \sqrt{T_e\varepsilon_0/(ne^2)}, \quad \lambda_{\rm Di} \sim \sqrt{T_i\varepsilon_0/(ne^2)}. \tag{3.46}$$

The conditions (3.45) are consistent with the condition $k\lambda_{\rm De} \gg 1$ assumed here, since $\lambda_{\rm De} \gg \lambda_{\rm Di}$.

We consider damping of ion-sound waves. Taking into account the imaginary part of the susceptibility $\operatorname{Im}\chi$, we have in place of (3.39)

$$\chi = -[\omega_{pi}/(\omega + i\operatorname{Im}\omega)]^2 + [C/(k\lambda_{\rm De})]^2 + i\operatorname{Im}\chi_e. \tag{3.47}$$

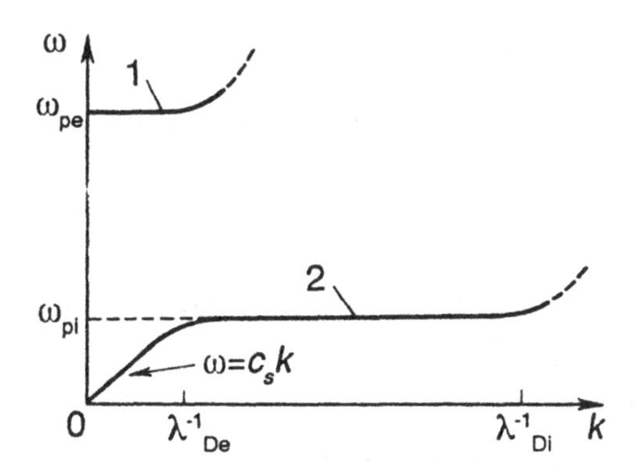

FIG. 1. Frequency spectra of waves in a plasma: (1) spectrum of electron plasma waves, (2) spectrum of ion-sound waves.

The ionic contribution to the damping is exponentially small and is neglected. Setting (3.47) to -1 and assuming the damping is weak, we find

$$\operatorname{Im}\omega \sim \frac{\omega^3}{\omega_{pi}^2}\operatorname{Im}\chi_e. \tag{3.48}$$

This is a general expression for the damping constant Im ω of ion-sound waves.

In the acoustic limit $k\lambda_{\mathrm{De}} \ll 1$ we have from (3.41) and (3.48)

$$\frac{Im\,\omega}{\omega} \sim (k\lambda_{\mathrm{De}})^2 \operatorname{Im}\chi_e. \tag{3.49}$$

Here, Im χ_e is given by (3.24). When $k\lambda_{\mathrm{De}} \ll 1$ it is less than unity, as noted above following the derivation of (3.24). Therefore the ratio Im $\omega/\omega \ll 1$ and the damping of ion-sound is weak. If $k\lambda_{\mathrm{De}} \gg 1$ then $\omega = \omega_{pi}$ and we obtain from (3.48)

$$\frac{Im\,\omega}{\omega} \sim \operatorname{Im}\chi_e. \tag{3.50}$$

According to (3.24), we have Im $\chi_e \ll 1$ when $k\lambda_{\mathrm{De}} \gg 1$ and therefore damping of longitudinal ion plasma waves is also small.

From the results obtained in this and the preceding sections, the following conclusions can be made about the types of waves in a plasma. For wave numbers in the region $k\lambda_{\mathrm{De}} < 1$ there are both electron plasma waves with frequency $\omega = \omega_{pe}$ as well as ion-sound waves with frequency $\omega = c_s k$. When $k\lambda_{\mathrm{De}} \gtrsim 1$ electron plasma waves disappear and the frequency of ion-sound waves is given by the general relation (3.40). Finally, in the region $k\lambda_{\mathrm{Di}} > 1$ ion-sound waves also disappear; in the region $\lambda_{\mathrm{De}}^{-1} \ll k \ll \lambda_{\mathrm{Di}}^{-1}$ they become ion plasma waves with frequency ω_{pi}. The frequency spectrum of waves in a plasma is shown in Fig. 1.

Problem 1. Show that the damping constant for ion-sound waves propagating in a Maxwellian plasma in the limit $k\lambda_{\mathrm{De}} \gg 1$ is of order

$$\operatorname{Im}\omega \sim \omega_{pi}\sqrt{m/M}(k\lambda_{\mathrm{De}})^{-3}.$$

Problem 2. Show that the damping constant for ion-sound waves propagating in a Maxwellian plasma in the limit $k\lambda_{\mathrm{De}} \ll 1$ is of order

$$\text{Im}\,\omega \sim \sqrt{m/M}\,\omega.$$

Problem 3. Show that for ion sound ($k\lambda_{\text{De}} \ll 1$) the explicit form of the interactions between electrons or between ions is not important.

3.5. Kinetic instabilities in a plasma

The electric field E for waves in a plasma with frequency $\omega(k)$ depends on time as

$$E \sim \exp[-i\omega(k)t]. \tag{3.51}$$

In the case of longitudinal waves the dependence of ω on k is found from the condition that the dielectric permittivity of the plasma must vanish, i.e., $\varepsilon(\omega,k) = 0$. We have determined the dependence of ω on k in the two preceding sections for the examples of plasma and ion-sound waves.

In general, the solution of the equation $\varepsilon(\omega,k) = 0$ is complex, i.e., $\omega = \text{Re}\,\omega + i\,\text{Im}\,\omega$. If $\text{Im}\,\omega < 0$, then the field (3.51) damps out in time (the example of Landau damping was considered above). If $\text{Im}\,\omega > 0$ then the electric field (3.51) grows exponentially in time, i.e., the plasma is unstable to such a process. In this case the quantity $\text{Im}\,\omega$ is called the *growth rate of instability.* In a real system the exponential growth of the field E is limited by nonlinear effects.

As already noted, in a collisionless plasma the imaginary part $\text{Im}\,\omega$ leads to Landau damping (see Sec. 3.2). For equilibrium we always have $\text{Im}\,\omega < 0$ and so the equilibrium state is stable against any perturbation in the plasma. This corresponds to the general principles of thermodynamics, according to which the entropy is a maximum for the equilibrium state of the system.

However, it is possible to have $\text{Im}\,\omega > 0$ for nonequilibrium states. This leads to an instability if any of the branches of plasma oscillations fall within this region of ω and k. Below, we consider two well-known examples of such instabilities.

First, we consider the so-called *beam instability.* Suppose a beam of electrons passes through a plasma which is at rest. Because of the directed motion of the beam, the system is not in equilibrium. We will assume that the concentration n_1 of electrons of the beam is small in comparison with the concentration n of electrons (or ions) of the plasma, i.e., $n_1 \ll n$. Let $v_{e1} = (T_{e1}/m)^{1/2}$ be the thermal velocity of the beam electrons (T_{e1} is the electron temperature of the beam), and let $v_e = (T_e/m)^{1/2}$ be the thermal velocity of the plasma electrons (T_e is the electron temperature of the plasma).

We assume that the macroscopic velocity V of the beam is large in comparison with the thermal velocity v_{e1} of the electrons in the beam: $V \gg v_{e1}$. We also assume that the frequency ω is large such that $\omega \gg kv_e, kv_{e1}$. In this limit the electric susceptibility of the system has the form [see (3.17)]:

$$\chi = -\left(\frac{\omega_{pe}}{\omega}\right)^2 - \left(\frac{\omega_{pe1}}{\omega_1}\right)^2. \tag{3.52}$$

Here $\omega_{pe} = [ne^2/(m\varepsilon_0)]^{1/2}$ is the plasma frequency of the plasma electrons [see (3.16)]. The second term in (3.52) corresponds to the contribution of the beam electrons. Indeed, the quantity $\omega_{pe1} = [n_1e^2/(m\varepsilon_0)]^{1/2}$ is the plasma oscillation frequency of the beam itself in the absence of the plasma. The quantity $\omega_1 = \omega$

$-kV$ is the oscillation frequency of the beam electrons, which is shifted in frequency relative to ω by kV because of the Doppler effect. This is just the frequency of the field sensed by an electron moving with velocity V ($V \gg v_e, v_{e1}$).

Because of its small concentration ($n_1 \ll n$) the beam exerts only a small perturbation on the plasma. Therefore the branch of the plasma frequency $\omega = \omega_{pe}$ in the absence of the beam is not changed significantly by the presence of the beam.

However, there is a completely different branch of plasma oscillations in which the frequency ω is close to kV and not to ω_{pe}. Setting (3.52) to -1, we find for this branch

$$\left(\frac{\omega_{pe}}{kV}\right)^2 + \left(\frac{\omega_{pe1}}{\omega - kV}\right)^2 = 1, \tag{3.53}$$

and hence

$$\omega = kV \pm \frac{\omega_{pe1}}{\sqrt{1 - (\omega_{pe}/kV)^2}}. \tag{3.54}$$

Because ω and kV are close, the frequency ω_1 is small in comparison with kV. According to (3.54), ω and kV will be close if kV is not too close to ω_{pe}. If kV is of the order of ω_{pe}, then $\omega_{pe1} \ll kV$ and the second term on the right-hand side of (3.54) is small in comparison with the first, i.e., ω is close to kV. The conditions $\omega \gg kv_e$, $\omega \gg kv_{e1}$ imposed above imply in this case that $V \gg v_{e1}, v_e$, which was assumed above.

When $kV > \omega_{pe}$ both roots in (3.54) are real and therefore damping or growth of the oscillations does not occur. If $kV < \omega_{pe}$, then the root in (3.54) for which $\operatorname{Im}\omega > 0$ leads to growth of the oscillations. Hence *the plasma is unstable to sufficiently long-wavelength oscillations.* This instability is called the beam instability.

Because $kV \lesssim \omega_{pe}$, it follows that $\omega \lesssim \omega_{pe}$ and the frequency of the new branch is within the frequency region of ordinary electron plasma waves. The condition $\omega \gg kv_e$ then implies that either $\omega_{pe} \gg kv_e$ or $k\lambda_{\mathrm{De}} \ll 1$, where λ_{De} is the electron Debye length [see (3.46)]. Hence *the energy of the electron beam in braking is transformed into plasma oscillations with wavelengths exceeding the electron Debye length.* This energy is then transformed into the other degrees of freedom in the plasma. The braking mechanism described above is much stronger than the braking of the electron beam due to collisions with plasma ions or electrons.

In addition to the imaginary part $\operatorname{Im}\omega$ arising when $k < \omega_{pe}/V$, as considered above, there is an exponentially small imaginary part $\operatorname{Im}\omega$ due to Landau damping. When Landau damping is taken into account, one of the roots of (3.54) has an imaginary part corresponding to growth of the oscillations for all values of k (although it is exponentially small).

We note that the beam interacts most strongly with waves whose phase velocity is equal to the velocity of the beam: $V = \omega/k$. Here, we can repeat the discussion of Sec. 3.2 [after Eq. (3.28)] on the stationary nature of the field of the wave as sensed by the electrons of the beam.

We consider another type of instability in which all of the electrons in the plasma move relative to the ions with velocity V. When $\omega \gg kv_e, kv_i$, where v_e is the average thermal velocity of the electrons and v_i is the average thermal velocity of the ions, i.e., $v_e = (T_e/m)^{1/2}$ and $v_i = (T_i/M)^{1/2}$, the susceptibility of the plasma is given by

(3.17), as in the case of the beam instability. Hence, in analogy with (3.53), the condition for longitudinal electric waves in the plasma has the form $\varepsilon = 0$:

$$\left(\frac{\omega_{pe}}{\omega - kV}\right)^2 + \left(\frac{\omega_{pi}}{\omega}\right)^2 = 1. \tag{3.55}$$

Here, ω_{pi} is the ion plasma frequency, which is equal to $[ne^2/(M\varepsilon_0)]^{1/2}$, or

$$\omega_{pi}^2 = \frac{m}{M}\Omega_e^2. \tag{3.56}$$

The contribution of the ions changes the dispersion relation for the plasma oscillations. In addition to the branch of electron plasma frequencies $\omega = \omega_{pe} + kV$, which exists in the absence of the contribution of the ions, there is a branch for which the wave number $k \sim -\omega_{pe}/V$ and the frequency ω is very small. Then (3.55) can be rewritten in the form

$$\left(\frac{\omega_{pe}}{kV}\right)^2 + \frac{m}{M}\left(\frac{\omega_{pe}}{\omega}\right)^2 = 1 \tag{3.57}$$

and hence we obtain for the frequency ω of the new branch

$$\pm\omega = \sqrt{\frac{m}{M}} \frac{\omega_{pe}}{\sqrt{1 - [\omega_{pe}/(kV)]^2}}. \tag{3.58}$$

Hence when $kV < \omega_{pe}$ the root in (3.58) for which $\operatorname{Im}\omega > 0$ leads to growth of the oscillations. The oscillation frequency in this case is purely imaginary. This type of oscillation is called the *Buneman instability.* According to (3.58), the electrons transfer their kinetic energy into ion plasma oscillations.

Problem 1. Show that in the case of the beam instability, at frequencies close to the electron plasma frequency, the growth rate of instability $\operatorname{Im}\omega$ is of order

$$\operatorname{Im}\omega \sim (n_1/n)^{1/3}\omega_{pe},$$

where n_1 is the concentration of electrons in the beam and n is the concentration of electrons in the plasma.

Problem 2. Show that in the case of the Buneman instability, when the frequency is close to the electron plasma frequency $\omega_{pe} + kV$, the growth rate of instability $\operatorname{Im}\omega$ is of order

$$\operatorname{Im}\omega \sim (m/M)^{1/3}\omega_{pe}.$$

Problem 3. Show that electron plasma oscillations with frequency ω_{pe} and electric-field amplitude E are unstable against decay into electron plasma oscillations with a lower frequency ω_{pe1} and ion sound with frequency $\omega_i = \omega_{pe} - \omega_{pe1}$, and show that the growth rate of instability is of order

$$\operatorname{Im}\omega \sim \frac{eE\sqrt{\omega_i}}{\sqrt{mT_e}\omega_{pe}}.$$

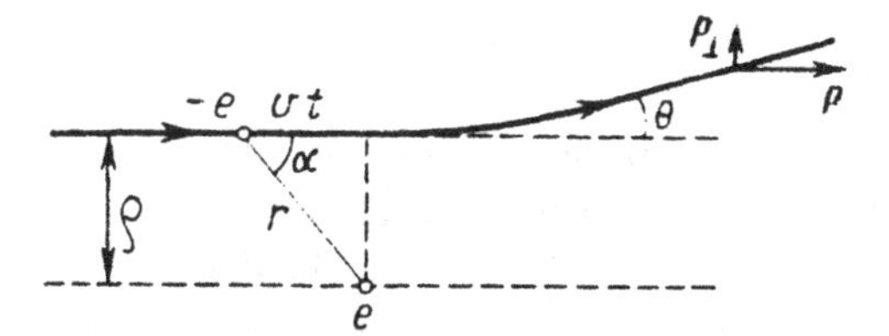

FIG. 2. Scattering in a Coulomb field by a small angle θ.

3.6. Collisions between charged particles in a plasma

The characteristic feature of a plasma is the long-range nature of the Coulomb interaction between the charged particles of the plasma. Hence in particle collisions an important role is played by large distances, when the particles are deflected only slightly and the momentum changes are small. The purpose of this section is to obtain the cross section for such collisions, as well as the mean free path and mean free time of particles in the plasma.

We estimate the differential cross section for scattering in the Coulomb field by small angles. The deflection angle $\theta = \overline{p_\perp}/\overline{p} \ll 1$, where $\overline{p}$ is the momentum of the particle, and $\overline{p}_\perp$ is the momentum acquired in the direction perpendicular to the initial direction:

$$p_\perp = \int_{-\infty}^{\infty} F_\perp \, dt = \int_{-\infty}^{\infty} \frac{e^2 \sin \alpha \, dt}{4\pi\epsilon_0 r^2}. \tag{3.59}$$

Here, $F_\perp$ is the force in the transverse direction and r is the distance between the colliding particles, i.e., $r = (\rho^2 + v^2t^2)^{1/2}$. The quantity ρ is the impact parameter of the collision and $v = p/\mu$ is the velocity of the particle (for small scattering angles the motion of the particle is practically uniform and rectilinear). Further, μ is the reduced mass of the colliding particles:

$$\mu^{-1} = M_1^{-1} + M_2^{-1}. \tag{3.60}$$

Finally, α is the angle between the direction of the Coulomb force and the direction of motion of the particle (see Fig. 2), hence $\sin \alpha = \rho/r$.

Evaluating the integral in (3.59) approximately, we find

$$p_\perp = \frac{e^2}{4\pi\varepsilon_0} \int_{-\infty}^{\infty} \frac{\rho \, dt}{(\rho^2 + v^2t^2)^{3/2}} \sim \frac{e^2}{\rho v \varepsilon_0}. \tag{3.61}$$

We then obtain an estimate for the deflection angle

$$\theta \sim e^2/(\mu v^2 \rho \varepsilon_0). \tag{3.62}$$

According to the definition of the differential cross section $d\sigma$ we have

$$d\sigma \sim \rho \, d\rho \sim d\left(\frac{e^2}{\mu v^2 \theta \varepsilon_0}\right)^2 \sim \left(\frac{e^2}{\mu v^2 \varepsilon_0}\right)^2 \frac{d\theta}{\theta^3}. \tag{3.63}$$

In a collision process we are normally interested in the momentum transfer between the colliding particles, which leads to the equilibrium state. The change of momentum along the direction of motion is $\overline{p}(1 - \cos \theta)$, since in the center-of-mass system of the colliding particles the momentum $\overline{p}$ changes only in direction, and not in magnitude. For small scattering angles θ the momentum transfer is of order $\bar{p}\theta^2$ and is therefore very small.

Obviously in a collision momentum is also transferred perpendicular to the motion of the particle. However, the transferred momentum in the transverse direction vanishes when averaged over direction in a plane perpendicular to the direction of motion of the particle. Therefore the momentum transfer is characterized not by the cross section $d\sigma$, but by the cross section $d\sigma_t = (1 - \cos\theta)d\sigma$, which is called the *cross section for momentum transfer* (see Problem 3 of Sec. 1.5). For small angles $\theta \ll 1$ we have $d\sigma_t \sim \theta^2 d\sigma$ and so $d\sigma_t \ll d\sigma$. Therefore for Coulomb interactions the difference between $d\sigma_t$ and $d\sigma$ is crucial, even from a qualitative point of view. For the case of collisions between neutral molecules considered in Chapters 1 and 2, which occur at small distances so that the typical deflection angle is $\theta \sim 1$, we have $d\sigma_t \sim d\sigma$, and therefore it is not necessary to introduce $d\sigma_t$ in a qualitative treatment.

Integrating (3.63) with respect to θ, we obtain an estimate for the cross section for momentum transfer in particle collisions in a plasma:

$$\sigma_t \sim \int \theta^2 \, d\sigma \sim \left(\frac{e^2}{\mu v^2 \varepsilon_0}\right)^2 \ln\frac{1}{\theta_0}. \tag{3.64}$$

Here, the angle θ_0 is the smallest angle for which the Coulomb nature of the force, which was assumed in the derivation, is correct. As we have seen above, screening of the Coulomb field takes place at impact parameters ρ of the order of the Debye length λ_{De} (see the introduction to this chapter). Therefore we obtain from (3.62)

$$\theta_0 \sim \frac{e^2}{\mu v^2 \lambda_{De} \varepsilon_0}. \tag{3.65}$$

We define the quantity

$$L = \ln\frac{1}{\theta_0} = \ln\frac{\mu\langle v\rangle^2 \lambda_{De}\varepsilon_0}{e^2} = \ln\frac{T\lambda_{De}\varepsilon_0}{e^2}, \tag{3.66}$$

where $\langle v\rangle = \sqrt{T/\mu}$ is the mean thermal velocity of the charged particle and T is the temperature of the plasma. The quantity L is called the *Coulomb logarithm*. We find from (3.64) and (3.66)

$$\sigma_t \sim \left(\frac{e^2}{T\varepsilon_0}\right)^2 L. \tag{3.67}$$

Because $\theta_0 \ll 1$, we have $L \gg 1$. The quantity θ_0 can be estimated using (3.2) for the Debye length:

$$\theta_0 \sim \frac{e^2}{T\varepsilon_0}\sqrt{\frac{ne^2}{\varepsilon_0 T}} \sim \left(\frac{e^2 n^{1/3}}{T\varepsilon_0}\right)^{3/2} \ll 1. \tag{3.68}$$

The condition $\theta_0 \ll 1$ implies the condition (3.1) for a rarefied plasma.

We have assumed that the scattering of particles in the plasma is classical and not quantum. This will be valid when the energy of a particle $\mu v^2/2$ is large in comparison with the typical quantum-mechanical interval $\hbar^2/\mu\rho^2$ between neighboring energy levels (see the introduction to this chapter). Hence we obtain a condition for classical scattering

$$\mu\rho v \gg \hbar. \tag{3.69}$$

We consider the condition for classical scattering for angles $\theta \sim 1$. According to (3.62), for angles $\theta \sim 1$ we have

$$\mu v^2 \sim e^2/(\rho\varepsilon_0). \tag{3.70}$$

This condition is equivalent to the **virial theorem:** *for the Coulomb interaction the total energy of the particle is of the order of its potential energy or the kinetic energy of relative motion.* Solving (3.70) for ρ and substituting the result into (3.69), we find the condition for classical scattering of the particles in the form

$$v \ll e^2/(\hbar\varepsilon_0). \tag{3.71}$$

This condition can also be expressed in terms of the plasma temperature by using the average thermal velocity of the particle for v in (3.71):

$$T \ll \mu e^4/(\hbar^2\varepsilon_0^2). \tag{3.72}$$

Equation (3.72) is not related to the condition (3.5) for a classical plasma. Condition (3.5) means that the plasma is a classical system when collisions between the particles are neglected, which has been assumed in this chapter up to now. If collisions are important then they can be treated classically (the particles move along specified trajectories during the scattering process) if the velocities of the particles are sufficiently small. Hence in addition to (3.5), the condition (3.72) must also be satisfied.

It is not difficult to show that if the condition for classical scattering is satisfied for scattering angles $\theta \sim 1$, it will also be satisfied for $\theta \ll 1$, corresponding to large impact parameters ρ. Hence (3.67) for the cross section for momentum transfer is correct when the condition (3.72) for classical scattering is satisfied.

The condition (3.72) is often realized in practice, since the right-hand side of (3.72) is a quantity of order 10^5 K in the case of the electron mass $\mu = m$. For ions it is much larger still; the corresponding temperature on the right-hand side of (3.72) is of order 10^8 K.

We consider now the mean free path l of charged particles in a plasma. From the general formula (1.1), it is given by

$$l \sim (n\sigma_t)^{-1} \tag{3.73}$$

(here we have replaced σ by σ_t). Recall that n is the concentration of electrons or ions in the plasma.

The quantity σ_t is determined from (3.67) by substituting the appropriate plasma temperature. For example, for electrons we find from (3.73) and (3.67)

$$l_e \sim \frac{(T_e\varepsilon_0)^2}{ne^4 L_e}, \tag{3.74}$$

where T_e is the electron temperature and L_e is the Coulomb logarithm for the electrons. According to (3.66) we have

$$L_e = \ln \frac{\lambda_{De} T_e \varepsilon_0}{e^2} \sim \ln \frac{T_e \varepsilon_0}{e^2 n^{1/3}}. \tag{3.75}$$

The mean free time of the electrons in the plasma is of order

$$\tau_e = \frac{l_e}{v_e} \sim \frac{T_e^{3/2} m^{1/2} \varepsilon_0^2}{n e^4 L_e}. \tag{3.76}$$

This quantity determines the time of relaxation of the electronic component of the plasma toward the equilibrium state.

We compare the mean free path of electrons in the plasma with their Debye length:

$$\frac{l_e}{\lambda_{De}} \sim \frac{T_e^2 \varepsilon_0^2}{n e^4 L_e} \sqrt{\frac{n e^2}{\varepsilon_0 T_e}} \sim \frac{1}{L_e} \left(\frac{T_e \varepsilon_0}{e^2 n^{1/3}} \right)^{3/2} \gg 1. \tag{3.77}$$

Because the plasma is rarefied [condition (3.1)], it follows from (3.77) that the mean free path of the electrons is large in comparison with the Debye length, in spite of the fact that $L_e \gg 1$ (the Coulomb logarithm is large only logarithmically and does not cancel out the effect of the term multiplying it). It has already been pointed out in the introduction to this chapter that the quantity λ_{De} cannot be identified with the mean free path.

From the inequality $l_e \gg \lambda_{De}$ we obtain $l_e/v_e \gg \lambda_{De}/v_e$ or $\tau_e \gg \omega_{pe}^{-1}$, i.e., *the collision frequency of the electrons* $\nu_e = \tau_e^{-1}$ *is small in comparison with the plasma frequency of the electrons* $\omega_{pe} = v_e/\lambda_{De}$ [*see* (3.16)].

We estimate the mean free path of the ions l_i with respect to ion-ion collisions. According to (3.67), the cross section for momentum transfer in ion-ion collisions is

$$\sigma_{ti} \sim \frac{e^4 L_i}{(\varepsilon_0^2 T_i)^2}. \tag{3.78}$$

Here, T_i is the temperature of the ions and L_i is the Coulomb logarithm for ion-ion collisions. In analogy with (3.75)

$$L_i = \ln \frac{T_i}{e^2 n^{1/3}} \gg 1. \tag{3.79}$$

When $T_i \sim T_e$ the Coulomb logarithms for electrons and ions are of the same order of magnitude.

From (3.73) and (3.78) we obtain

$$l_i \sim \frac{1}{n \sigma_{ti}} \sim \frac{1}{n L_i} \left(\frac{\varepsilon_0 T_i}{e^2} \right)^2. \tag{3.80}$$

The mean free time τ_i of the ions in ion-ion collisions is

$$\tau_i = l_i / v_i \sim T_i^{3/2} M^{1/2} \varepsilon_0^2 / (n e^4 L_i). \tag{3.81}$$

Comparing (3.81) with (3.74), assuming $T_e \sim T_i$, we obtain

$$\tau_e / \tau_i \sim \sqrt{m/M} \ll 1, \tag{3.82}$$

and so *the electronic component of the plasma reaches equilibrium much more rapidly than the ionic component.*

The quantity τ_e characterizes the time of momentum transfer from electrons to other electrons, and also from electrons to ions, since in electron-ion collisions the reduced mass μ is equal to the mass of the electron m. However, τ_e is in no way related to the time of energy exchange between electrons and ions. Indeed, in electron-ion collisions the change in the energy of an electron (assuming $T_i \sim T_e$) is of order [see (1.46)]

$$\Delta\varepsilon \sim \sqrt{\frac{m}{M}}\, T_e \ll T_e, \tag{3.83}$$

and hence is small in comparison with the energy of the electron itself, whereas the momentum of the electron changes by a quantity of the order of the momentum itself in each collision with an ion (because the change in direction of the momentum can be large). From the discussion at the end of Sec. 1.5, the energy of the electron changes by a quantity of the order of the energy T_e itself only after M/m collisions. Therefore the time τ_{ei} required to establish equilibrium between the electrons and the ions is of order

$$\tau_{ei} \sim \frac{M}{m}\, \tau_e. \tag{3.84}$$

Using (3.76), we obtain from (3.84)

$$\tau_{ei} \sim T_e^{3/2} M \varepsilon_0^2 / (n e^4 m^{1/2} L_e). \tag{3.85}$$

We see from (3.85) not only is τ_{ei} large in comparison with τ_e, but it is also large in comparison with τ_i [see (3.81)]. Hence *the transfer of energy between electrons and ions and the corresponding equalization of the electron and ion temperatures is a much slower process than the approach of either the electronic or ionic components to equilibrium separately.* From the definition of τ_{ei} as a relaxation time, we can write

$$\frac{d}{dt}(T_e - T_i) = -(T_e - T_i)/\tau_{ei}. \tag{3.86}$$

Hence

$$|T_e - T_i| \sim T_e \exp(-t/\tau_{ei}). \tag{3.87}$$

We conclude this section by discussing the question of when collisions in the plasma are significant and when they are not. The answer to this question depends on the type of perturbation in the plasma. Consider again the perturbation of the plasma produced by a variable electric field with frequency ω. For collisions to be negligible the collision frequency $\nu = \tau^{-1}$ must be small in comparison with the frequency of the field ω. The smallest of the mean free times is τ_e. Therefore we obtain from (3.76) the condition for neglecting collisions in the form

$$T_e^{3/2} m^{1/2} \omega \varepsilon_0^2 \gg n e^4 L_e. \tag{3.88}$$

An analogous condition can be imposed on the mean free path of the particles in the plasma: it must be large in comparison with the wavelength $\lambda = k^{-1}$ of the variable electric field. From (3.74) we obtain a lower bound on the wave number:

$$kT_e^2\varepsilon_0^2 \gg ne^4L_e. \tag{3.89}$$

If $\omega < kv$ then (3.88) must be satisfied; otherwise, (3.89) must be satisfied.

Problem 1. Using the Born approximation of quantum mechanics show that in the limiting case $v \gg e^2/\hbar\varepsilon_0$ [the opposite of the condition (3.71) for classical scattering] the Coulomb logarithm has the form

$$L = \ln\frac{\mu\lambda_{De}v}{\hbar}.$$

Problem 2. Show that in a weakly ionized gas it is possible to introduce separate electron T_e and ion T_i temperatures if

$$n_e \gg n_a \frac{m}{M}\frac{\sigma T_e^2\varepsilon_0^2}{e^4L_e},$$

where n_e and n_a are the concentrations of electrons (ions) and atoms, and σ is the cross section for electron-atom collisions.

3.7. Thermal conductivity, viscosity, and electrical conductivity of a plasma

Using the expressions obtained above for the mean free path of charged particles in a plasma and the general expressions for the kinetic coefficients obtained in Chaps. 1 and 2, we estimate the kinetic coefficients for plasmas.

The thermal conductivity λ of a plasma is, according to (1.34),

$$\lambda \sim nvl. \tag{3.90}$$

λ is determined mainly by electrons, since their velocities v are larger, while according to (3.74), the mean free paths l_e and l_i are of the same order of magnitude when the electron and ion temperatures are comparable. We obtain

$$\lambda \sim \frac{\varepsilon_0^2 T_e^{5/2}}{e^4\sqrt{m}L_e}. \tag{3.91}$$

We see from (3.91) that the contribution of the ions to the thermal conductivity is smaller than that of electrons by the factor $\sqrt{M/m}$, assuming $T_i \sim T_e$.

We turn now to the viscosity of a plasma. We have from (1.38)

$$\eta \sim \sqrt{MT}/\sigma_t. \tag{3.92}$$

It follows from this equation that, in contrast to the thermal conductivity, the viscosity is determined by the ions. Substituting (3.78) for the cross section for momentum transfer in ion-ion collisions into (3.92), we find

$$\eta \sim \frac{\varepsilon_0^2 M^{1/2} T_i^{5/2}}{e^4 L_i}. \tag{3.93}$$

For ion-electron collisions M in (3.93) is replaced by the reduced mass of the electron and ion, i.e., m, and so electron-ion collisions (as well as electron-electron collisions) play only a small role.

Finally, we estimate the electrical conductivity of the plasma σ_e. We set $c \sim 1$ in the formula (2.5) and replace σ by σ_t. Then

$$\sigma_e \sim \frac{e^2}{\sigma_t \sqrt{mT}}. \tag{3.94}$$

We see that the conductivity is due to the electrons, since according to (3.67) the cross sections for momentum transfer for electrons and ions are comparable to one another when $T_e \sim T_i$. Substituting (3.67) into (3.94), we obtain an estimate for the conductivity of the plasma:

$$\sigma_e \sim \frac{\varepsilon_0^2 T_e^{3/2}}{e^2 m^{1/2} L_e}. \tag{3.95}$$

Note that in calculating the electronic contribution to the kinetic coefficients it is usually necessary to include both electron-electron and electron-ion collisions; both contributions are of the same order of magnitude, except in special cases.

Problem 1. Show that in the high-frequency limit $\omega \gg kv$ the damping time of plasma waves due to collisions is of order of the time τ_e given by (3.76).

Problem 2. A temperature gradient dT/dx created in the plasma leads to an electric field $E = \alpha\, dT/dx$. The quantity α is called the *thermoelectric coefficient*. Show that it is of order e^{-1}, where e is the charge of the electron.

Problem 3. Show that the equilibrium temperature is established in a plasma much more rapidly than the equilibrium density.

Problem 4. Show that the conductivity of a plasma is determined by ion-electron collisions, but not electron-electron collisions because of conservation of the total momentum of the electronic component.

Problem 5. Show that even when the degree of ionization of a gas is weak the conductivity is determined by collisions of electrons with ions and not with neutral atoms.

Chapter 4

Dielectrics

In this chapter we consider the kinetic properties of dielectric crystals using qualitative methods. The term dielectric implies an absence of free electrons in the solid. In the introduction we consider the properties of dielectrics in equilibrium at a given temperature; the results obtained here will be used in determining the kinetic characteristics of nonequilibrium processes in the different sections of this chapter.

The characteristic feature of a crystal is that the atoms perform small vibrations about their equilibrium positions (the lattice sites of the crystal). These vibrations represent the equilibrium thermal motion of the atoms (or ions) at a given temperature T. In the first approximation the vibrations can be assumed to be harmonic.

We estimate the vibration amplitude ξ_T. It is assumed that the vibrations are classical (the conditions under which thermal vibrations of atoms in a crystal can be treated as classical will be discussed below). Then the equipartition theorem holds, which states that the vibrational energy of an atom is of the order of the temperature T,

$$M\omega_0^2\xi_T^2 \sim T. \tag{4.1}$$

Here, M is the mass of an atom and ω_0 is its natural frequency of vibration. Hence we obtain

$$\xi_T \sim \sqrt{T/(M\omega_0^2)}. \tag{4.2}$$

We give a very rough estimate of the vibration frequency ω_0. The potential energy between neighboring ions of the crystal is of order $e^2/(d\varepsilon_0)$, where e is the charge of an ion and d is the distance between neighboring ions of the crystal (the so-called *lattice constant*. When d changes by a small quantity ξ (the deviation of the ion from the equilibrium position) the potential energy becomes $e^2/[(d + \xi)\varepsilon_0]$. Expanding this quantity in a power series in ξ, and using the fact that the linear term in ξ drops out because the force on an ion in the equilibrium position is zero, the lowest-order term in ξ is the quadratic term, which is of order $e^2\xi^2/(d^3\varepsilon_0)$. Setting this quantity equal to the vibrational energy of a harmonic oscillator $M\omega_0^2\xi^2$, we obtain an order-of-magnitude estimate for the characteristic vibration frequency ω_0 of an ion about its equilibrium position:

$$\omega_0 \sim \sqrt{e^2/(Md^3\varepsilon_0)} \sim \sqrt{ne^2/(M\varepsilon_0)}. \tag{4.3}$$

Here, $n \sim d^{-3}$ is the concentration of ions in the crystal. A similar formula holds for neutral atoms, since in a crystal the separation of neighboring atoms is small and there is significant overlap of the electron shells. Therefore the interactions between atoms are described very roughly by the static Coulomb interaction.

Note that (4.3) has the same form as (3.38) for the frequency of ion plasma oscillations in a rarefied plasma. The difference is that the frequency ω_0 in a crystal is much higher than the plasma frequency in a rarefied plasma because of the much larger value of n in a solid.

Substituting (4.3) into (4.2), we find

$$\xi_T \sim \sqrt{Td\varepsilon_0/e^2}d. \tag{4.4}$$

The assumption of harmonic oscillations is valid when $\xi_T \ll d$. From (4.4) this condition becomes

$$T \ll e^2/(d\varepsilon_0) \text{ or } T \ll e^2 n^{1/3}/\varepsilon_0. \tag{4.5}$$

We note that this is the opposite of the condition (3.1) for a rarefied plasma. At room temperature (4.5) is easily satisfied, since $T \sim 10^{-2}$ eV, while $e^2/(d\varepsilon_0)$ is of order 10 eV.

In a crystal lattice the atoms do not vibrate independently of one another. Because of the strong interactions between atoms, the vibrations of the atoms are coupled and therefore all of the atoms of the crystal vibrate with the same frequency ω_0.

We discuss the condition under which classical mechanics can be used to describe the vibrations of atoms in a crystal lattice. In quantum mechanics a wave with frequency ω_0 corresponds to a particle called a *phonon* (wave-particle duality). The frequency ω_0 and energy ε of a phonon are related by $\varepsilon = \hbar\omega_0$, where $\hbar$ is Planck's constant. Therefore the number of phonons per atom of the crystal is roughly $T/(\hbar\omega_0)$, since T is the thermal energy of an atom. If the number of phonons is large, i.e., $T \gg \hbar\omega_0$, then the discrete nature of the phonons is not important and the vibrations can be treated using classical mechanics. Using (4.3), the condition under which classical mechanics can be used to treat the thermal vibrations of atoms in a crystal lattice is

$$T \gg \hbar\sqrt{ne^2/(M\varepsilon_0)} = \Theta. \tag{4.6}$$

The quantity Θ is called the *Debye temperature*. It is of order 100–500 K for most solids. Even at temperatures $T \sim \Theta$ classical mechanics is still a good approximation; it is only when $T \ll \Theta$ that quantum effects begin to be felt.

4.1. Thermal expansion of a crystal

Thermal expansion of a crystal is an equilibrium process. Nevertheless, before turning to nonequilibrium processes, it will be useful to consider it as an example of the application of qualitative methods to the vibrations of atoms in a crystal lattice.

In the harmonic approximation the average displacement $\langle\xi\rangle$ of an atom from its lattice site is equal to zero and therefore thermal expansion of the crystal, i.e., a change of its average volume with temperature, cannot occur. The quantity $\langle\xi\rangle$

becomes nonzero when the anharmonic nature of the vibrations is taken into account. We will assume the vibrations are classical.

Expanding the potential energy of an atom in a power series in the deviation ξ from the equilibrium position $\xi=0$ (see the introduction to this chapter), we obtain

$$U=\tfrac{1}{2}M\omega_0^2\xi^2+g\xi^3. \tag{4.7}$$

The second term in (4.7) is the anharmonic part of the potential energy in the lowest (cubic) approximation. The vibrations in the potential (4.7) are stable at small values of $\xi \ll d$.

The quantity g can be estimated as follows. The expansion (4.7) becomes meaningless when $\xi \sim d$, when the anharmonic term becomes comparable to the harmonic term, since when $\xi \sim d$ the concept of vibration itself becomes meaningless. Setting $\xi \approx d$ in (4.7) and setting the two terms on the right-hand side of (4.7) equal to one another, we find

$$g \sim M\omega_0^2/d \sim e^2/(d^4\varepsilon_0). \tag{4.8}$$

Here, we have used (4.3) for the frequency ω_0.

The probability of a given value ξ at temperature T is determined by the equilibrium Boltzmann distribution

$$w(\xi) \sim \exp[-U(\xi)/T]. \tag{4.9}$$

This result is correct if we assume that the different vibrations are independent of one another (the analog of an ideal gas). Substituting (4.8) into (4.9) and expanding the exponential in a power series in the term $g\xi^3$, we obtain

$$w(\xi) \sim [1-g\xi^3/T]\exp[-M\omega_0^2\xi^2/(2T)]. \tag{4.10}$$

The anharmonic correction in (4.10) is small in comparison with unity, which justifies including only the first term of the expansion. Indeed, setting $\xi \sim \xi_T$ and using (4.4), we find

$$g\xi_T^3/T \sim (Td\varepsilon_0/e^2)^{1/2} \ll 1. \tag{4.11}$$

Let $\langle\xi\rangle$ be the average equilibrium displacement of an atom. It is given by the usual formula of probability theory for the average value:

$$\langle\xi\rangle = \int_{-\infty}^{\infty} \xi w(\xi)d\xi \Big/ \int_{-\infty}^{\infty} w(\xi)d\xi. \tag{4.12}$$

Substituting (4.10) into (4.12), we find that the first term in (4.10) leads to a zero value of the numerator of (4.12), since the integrand is an odd function. Therefore

$$\langle\xi\rangle = -\frac{g}{T}\frac{\int \xi^4 w(\xi)d\xi}{\int w(\xi)d\xi} \sim \frac{g}{T}\xi_T^4. \tag{4.13}$$

Substituting (4.4) for ξ_T into (4.13), we find

$$\langle\xi\rangle \sim \frac{Td\varepsilon_0}{e^2}d \ll \xi_T \sim \sqrt{\frac{Td\varepsilon_0}{e^2}}\,d. \tag{4.14}$$

As expected, we have $\langle\xi\rangle \ll \xi_T$, i.e., the anharmonic displacement of an atom from its equilibrium position is small in comparison with the amplitude of its thermal vibrations (and therefore is small in comparison with the distance between nearest-neighbor atoms on the crystal lattice).

The *coefficient of linear expansion* α is defined as the relative expansion of the crystal when it is heated one degree Kelvin

$$\alpha=\frac{1}{d}\frac{d\langle\xi\rangle}{dT}. \tag{4.15}$$

The quantity d appears in this formula because the displacement $\langle\xi\rangle$ of a single atom occurs over a distance equal to the lattice constant d; for N atoms the total displacement, and therefore the total length of the expansion, is larger by a factor of N. Substituting (4.14) into (4.15), we find

$$\alpha \sim \varepsilon_0 d/e^2. \tag{4.16}$$

Hence *in the classical approximation the coefficient of linear expansion of a crystal does not depend on its temperature.*

Problem 1. Show that at low temperatures $T \ll \Theta$ the coefficient of linear expansion is proportional to the cube of the temperature (in the case of a dielectric).

Problem 2. Show that at arbitrary temperatures $T \lessgtr \Theta$ the ratio of the coefficient of linear expansion to the heat capacity of a crystal is independent of temperature (*Grüneisen's law*).

Problem 3. Using (5.3) show that at very low temperatures the coefficient of linear expansion of a metal is proportional to the first power of the temperature.

4.2. Sound waves in a crystal

In the introduction to this chapter we considered equilibrium vibrations of atoms in a crystal lattice at a given temperature. Here, we discuss the excitation of mechanical vibrations of the lattice by an external force, which disturbs the system from the state of thermal equilibrium. The amplitude of the vibrations depends on the strength of the external force. The wavelength of the vibrations can vary from the lattice constant d to infinity (the wavelength λ cannot be smaller than the lattice constant d because the structure of the crystal is discrete). We will assume that $\lambda \gg d$. For simplicity we assume a one-dimensional chain of identical atoms separated by the distance d (Fig. 3). In addition, we assume that the atoms interact only with their nearest neighbors (the interaction with more distant neighbors falls off rapidly because of screening). Let ξ_n be the displacement of atom n along the one-dimensional chain.

From Newton's second law, we obtain the equation of motion for atom n

$$M\ddot{\xi}_n=C(\xi_{n+1}-\xi_n)+C(\xi_{n-1}-\xi_n). \tag{4.17}$$

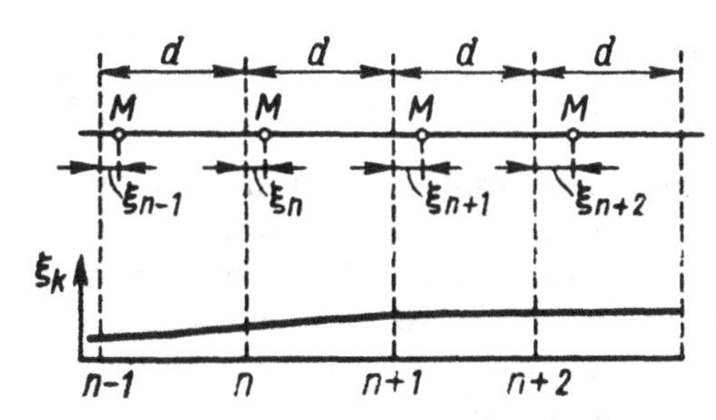

FIG. 3. One-dimensional chain representing a crystal lattice. The lower part of the figure shows the displacement of atoms in the presence of a propagating wave, where n is the number of the lattice point.

Here, C is the elastic constant in the elastic force $F = -C\xi$ acting on the atom. Because $F = -\partial U/\partial\xi$, where U is the potential energy of vibration, we have $F = -M\omega^2\xi$ and therefore $C = M\omega^2$. Using (4.3), we obtain an estimate for the elastic constant C

$$C \sim \frac{e^2}{d^3\varepsilon_0}. \tag{4.18}$$

Assuming that the amplitudes ξ_n of vibration of the neighboring atoms differ only slightly from one another, which is true when $\lambda \gg d$, we can write

$$\xi_{n\pm 1} - \xi_n = \pm\frac{d\xi_n}{dx}d + \frac{1}{2}\frac{d^2\xi_n}{dx^2}d^2. \tag{4.19}$$

Substituting (4.19) into (4.17), we obtain the equation for the vibrations

$$M\frac{\partial^2\xi}{\partial t^2} = Cd^2\frac{\partial^2\xi}{\partial x^2}, \tag{4.20}$$

where the index n can now be dropped.

The wave equation (4.20) describes harmonic natural vibrations of the quantity $\xi(x,t)$:

$$\xi(x,t) \sim \exp[i(kx - \omega t)]. \tag{4.21}$$

Here, ω is the frequency of vibration and k is the wave number ($\lambda = k^{-1}$ is the wavelength). Substituting (4.21) into (4.20), we obtain a relation between ω and k:

$$\omega = c_s k, \tag{4.22}$$

where

$$c_s \sim (Cd^2/M)^{1/2}. \tag{4.23}$$

Vibrations of this type are called *acoustic vibrations* and the quantity c_s is just the speed of sound. A similar dependence was discussed in Sec. 3.4 for ion sound in a plasma [see (3.41)]. We have seen that in a plasma (and also in a rarefied neutral gas) the speed of sound c_s is of order $\sqrt{T/M}$ [see (3.42)]. Substituting the estimate (4.18) for the elastic constant into (4.23), we obtain an estimate for the speed of sound in a solid:

$$c_s \sim \sqrt{\frac{e^2}{Md\varepsilon_0}}. \tag{4.24}$$

It follows from the condition (4.5) that this quantity is much larger than $\sqrt{T/M}$, i.e., *the speed of sound in a solid is much larger than in a gas or plasma.* In most solids it is about 5–10 km/s, which is consistent with (4.24). Vibrations with the dispersion law (4.22) are called *acoustic vibrations.* As mentioned above, the smallest possible wavelength of an acoustic vibration is of the order of the lattice constant d. Therefore the highest possible frequency is

$$\omega_0 \sim \sqrt{e^2/(Md^3\varepsilon_0)}, \tag{4.25}$$

and hence it is of the order of the quantity (4.3), which was found in a very rough approximation assuming the vibrations of the atoms are independent of one another. When the coupling between the vibrations is taken into account, we obtain the acoustic spectrum (4.22).

Problem 1. Show that in a one-dimensional chain of atoms with nearest-neighbor interactions, the dependence of the vibration frequency ω on the wave number k for arbitrary k has the form

$$\omega \sim \sqrt{C/M}\,\sin(kd/2).$$

Problem 2. Show that in a one-dimensional alternating chain of two kinds of atoms there are two types of vibrations: acoustic vibrations and another type of vibration in which the different atoms move out of phase with respect to one another and the vibration frequency is given by (4.3) and depends only weakly on k. Vibrations of this type are called *optical vibrations.*

4.3. Thermal conductivity of a dielectric

We consider the process of heat transfer in a dielectric crystal in the presence of a small temperature gradient dT/dx. Our goal is to estimate the thermal conductivity of the dielectric. We limit ourselves to classical vibrations, assuming that the temperature of the dielectric is not too low in comparison with the Debye temperature (see the introduction to this chapter).

Let T be the vibrational energy of an atom and n be the concentration of atoms in the dielectric (the number of atoms per unit volume). Then nT is the energy of thermal motion per unit volume of the dielectric. This corresponds to the classical law of Dulong and Petit, which is a consequence of the classical equipartition theorem. Acoustic vibrations propagating in the direction of the temperature gradient are the carriers of thermal energy. At high temperatures the flux density of thermal energy is of order

$$q \sim nTc_s. \tag{4.26}$$

When the temperature is low in comparison with the Debye temperature we must replace (4.26) by $q \sim CTc_s$, where C is the heat capacity per unit volume of the dielectric.

A temperature gradient disturbs the system from the state of thermal equilibrium. The resulting heat flux q can be found from (4.26) following the same reasoning used to obtain (1.5) and (1.33) for the heat flux in a neutral gas:

$$q \sim -nc_s l \frac{dT}{dx}. \tag{4.27}$$

Here, l is the mean free path of the sound waves. Note that the velocity c_s takes the place of the velocity v of thermal motion of the molecules in (1.33). But in a gas this is the same order of magnitude as the speed of sound. Hence there is a complete analogy between (1.33) and (4.27). From the definition (1.31) we therefore obtain for the thermal conductivity λ

$$\lambda \sim nc_s l, \tag{4.28}$$

which is analogous to the estimate (1.34) in the case of a neutral gas.

We estimate the mean free path l of a sound wave. In the harmonic approximation sound waves can propagate without bound and without interacting with one another. Therefore to calculate l it is essential to take into account the anharmonicity of the vibrations in the crystal. This has already been done in Sec. 4.1.

The equilibrium displacement of an atom $\langle \xi \rangle$ given by (4.14) occurs over a distance d. The displacement of the lattice by the lattice constant d accumulates over N atoms, such that $d \sim N\langle \xi \rangle$. Therefore we obtain from (4.14)

$$N \sim d/\langle \xi \rangle \sim e^2/(dT\varepsilon_0) \gg 1. \tag{4.29}$$

The corresponding length

$$l \sim Nd \sim e^2/(\varepsilon_0 T) \gg d \tag{4.30}$$

can be interpreted as the mean free path of a sound wave, since after traveling a distance l the sound wave does not encounter an atom of the crystal in the correct place (for an ideal crystal) and hence it is scattered. We see from (4.30) that the mean free path of a sound wave is large in comparison with the lattice constant. It must be small in comparison with the wavelength (see the next section) in order for the system to be considered as slightly nonuniform: $d \ll l \ll \lambda$.

In lattices consisting of two or more types of atoms optical vibrations are also present. The frequency of optical vibrations depends only weakly on the wave number (see Problem 2 of the preceding section). Because of this weak dependence the group velocity $d\omega/dk$ of these waves is small and therefore heat is carried by acoustic vibrations and not optical vibrations. Recall that energy is transferred with the group velocity of the wave $d\omega/dk$, and not the phase velocity ω/k.

It follows from (4.30) that the collision frequency $\nu = c_s/l$ is of order

$$\nu \sim \frac{\sqrt{\varepsilon_0} T}{e\sqrt{Md}}. \tag{4.31}$$

It is proportional to the temperature T. This dependence can also be understood in the language of phonons: the collision frequency of a given phonon with other phonons is proportional to the number of phonons. But we have seen that the number of phonons is of order $T/(\hbar\omega_0)$, which establishes the temperature dependence of the collision frequency.

Other important factors determining the mean free path l of a sound wave are the finite size of the crystal and scattering of waves by impurity centers and other deviations from an ideal crystal lattice.

Substituting (4.30) into (4.28) and using the estimate (4.24) for the speed of sound, we obtain a final estimate for the thermal conductivity of a dielectric:

$$\lambda \sim ne^3/(T\sqrt{Md\varepsilon_0^3}). \tag{4.32}$$

We see that λ is inversely proportional to the temperature. In a neutral gas, on the other hand, λ is proportional to $\sqrt{T}$ [see (1.34)].

From (4.30) we conclude that l increases with decreasing temperature and when it becomes larger than the sample size L it is necessary to replace l by L in (4.28) (see the analogous procedure in Sec. 1.7 for a highly rarefied gas). We then obtain

$$\lambda \sim nc_s L. \tag{4.33}$$

It is evident that the temperature dependence of the thermal conductivity disappears.

We compare the thermal conductivities of a dielectric and a gas at identical temperatures. Dividing (4.32) by (1.34), we find

$$\frac{\lambda_d}{\lambda_g} \sim \frac{ne^3}{T\sqrt{Md\varepsilon_0^3}} \frac{\sigma\sqrt{M}}{\sqrt{T}} \sim \left(\frac{e^2}{dT\varepsilon_0}\right)^{3/2} \gg 1. \tag{4.34}$$

Here, we have used the estimate $\sigma \sim d^2$ for the geometric cross section for collision of neutral molecules with one another in a gas: the linear dimension of a molecule is the same order of magnitude as the lattice constant in a crystal d. We see from (4.34) that *the thermal conductivity of a solid is much larger than that of a gas.* According to (4.34), the ratio of the thermal conductivities is of order 10^4. This is consistent with typical experimental data.

Problem 1. Show that for the phonon gas in a dielectric, the kinematic viscosity ν at high temperatures $T > \Theta$ is approximately

$$\nu \sim lc_s \sim \frac{e^3}{T\sqrt{Md\varepsilon_0^3}}. \tag{4.35}$$

Problem 2. Show that at low temperatures $T \ll \Theta$ the part of the thermal conductivity of a dielectric due to collisions between acoustic phonons is inversely proportional to the square of the temperature.

Problem 3. Show that the part of the thermal conductivity of a dielectric due to scattering of acoustic phonons by impurity centers is inversely proportional to the first power of the temperature.

4.4. Sound absorption in a dielectric

In this section we consider sound absorption in a dielectric due to dissipation of acoustic energy by means of heat conduction in the crystal. A sound wave of the form (4.21) propagates in the crystal with a certain frequency ω, wavelength $\lambda = k^{-1}$, and amplitude. The wavelength is assumed to be much larger than the mean free path (4.30). Then the process can be described macroscopically (a slightly nonuniform gas in the terminology of Chap. 1, where the wavelength λ plays the role of the linear dimension L of the nonuniformity). Hence we assume $kl \ll 1$ (which implies that $kd \ll 1$).

We consider first the energy flux density in a sound wave. The vibrational energy of an atom is roughly $M\omega^2\xi^2$, where ξ is the displacement of the atom (4.21) from its equilibrium position, M is the mass of the atom, and ω is the frequency of vibration. The energy flux density is obtained by multiplying this expression by the concentration of atoms n and by the energy propagation velocity, i.e., the speed of sound c_s:

$$W \sim nc_s M\omega^2\xi^2. \tag{4.36}$$

The sound wave induces a temperature gradient

$$\frac{dT}{dx} = \frac{dT}{d\xi}\frac{d\xi}{dx}. \tag{4.37}$$

We estimate this gradient as follows. According to (4.21), the derivative $d\xi/dx$ is approximately $k\xi$. The derivative $dT/d\xi$ is estimated using the fact that the characteristic length over which the temperature varies in the process of propagation of a sound wave is the wavelength $\lambda = k^{-1}$, which plays the role of the linear dimension of the nonuniformity in the system (see Sec. 1.3). Hence the derivative is approximately

$$\frac{dT}{d\xi} \sim \frac{T}{\lambda} \sim kT. \tag{4.38}$$

Therefore using (4.37) and (4.38), we obtain the following estimate for the temperature gradient:

$$\frac{dT}{dx} \sim k^2\xi T. \tag{4.39}$$

We now consider the quantity of heat Q absorbed in a unit volume of dielectric per unit time. The heat flux density $q = -\lambda dT/dx$ determines the quantity of heat passing through a unit area of dielectric perpendicular to the x axis per unit time. Similarly, we have for position $x + dx$

$$q + dq = -\lambda(T + dT)\frac{dT}{dx}. \tag{4.40}$$

Subtracting the heat flux density at point x from (4.40), we obtain the change of the heat flux over a length dx:

$$dq = \frac{d\lambda}{dT} dT \frac{dT}{dx} = \frac{d\lambda}{dT} \left(\frac{dT}{dx} \right)^2 dx. \tag{4.41}$$

The quantity of heat absorbed per unit volume of the dielectric per unit time is obtained from (4.41) by setting $dx = 1$:

$$Q \sim \frac{\lambda}{T} \left(\frac{dT}{dx} \right)^2. \tag{4.42}$$

Substituting (4.39) into (4.42), we obtain an estimate for the absorbed energy:

$$Q \sim \lambda k^4 \xi^2 T. \tag{4.43}$$

The *absorption coefficient of sound* is defined as the density Q of absorbed acoustic energy per unit time divided by the flux density W of incident acoustic energy: $\alpha_s = Q/W$. This coefficient has the dimensions of an inverse length, i.e., α_s^{-1} characterizes the distance over which absorption of sound becomes significant. Dividing (4.43) by (4.36), we find the absorption coefficient of sound in a dielectric:

$$\alpha_s = \frac{Q}{W} \sim \frac{\lambda k^2 T}{n M c_s^2}. \tag{4.44}$$

Substituting the estimate (4.32) for the thermal conductivity λ into (4.44), we finally obtain

$$\alpha_s \sim k^2 d \sim \varepsilon_0 \omega^2 M d^2 / e^2. \tag{4.45}$$

As expected, the arbitrary displacement ξ drops out of the expression for α_s (Q and W are both proportional to ξ^2).

It follows from (4.45) that *the absorption coefficient of sound in a dielectric is proportional to the square of the frequency and does not depend on temperature.* As is evident from the derivation, this statement assumes classical vibrations, i.e., temperatures $T \gg \Theta$, where Θ is the Debye temperature.

Problem 1. Show that the contribution of viscosity of the phonon gas to the absorption coefficient of sound is of the same order of magnitude as the contribution of the thermal conductivity when $T \gg \Theta$.

Problem 2. Show that at low temperatures $T \ll \Theta$ the viscous part of the absorption coefficient of sound in a dielectric is proportional to the square of the frequency and inversely proportional to the temperature.

Problem 3. Show that in the opposite limiting case of short wavelength, when $kl \gg 1$ (l is the mean free path of the phonons) the attenuation of a sound wave can be treated as the result of absorption of acoustic phonons in collisions with thermal phonons. Show that the absorption coefficient of sound is proportional to the frequency and to the fourth power of the temperature. Discuss the correspondence between this problem and the mechanism of Landau damping (Sec. 3.2) and show that the dominant role is played by phonons moving in phase with the sound wave.

Problem 4. Show that at low temperatures $T \ll \Theta$ the dominant role in sound absorption is played by acoustic phonons with energy $\hbar\omega \sim T$.

Problem 5. Show that the absorption coefficient of sound per unit time differs by a constant factor (the speed of sound) from the absorption coefficient per unit length.

Chapter 5

Metals

In this chapter we consider the kinetic properties of metals, in which there are conduction electrons, unlike dielectrics. The conduction electrons can transport electric charge. However, they collide with one another, with thermal sound waves in the crystal lattice, and with impurity atoms (or ions) in the crystal. The electrons do not collide with the lattice ions located exactly at the lattice sites in the absence of thermal vibrations. Indeed, we will see below that the electrons in a metal are fundamentally quantum particles, not classical particles. The electrons, treated as de Broglie waves, interfere with one another when scattered from an ideal lattice such that the rescattered wave is identical to the incident wave. The thermal vibrations of the ions about the lattice sites are random and they therefore destroy the coherence of the superposition of the interfering electron waves, and this is the mechanism leading to actual scattering of electrons.

An electron in a metal (unlike an ion) is a fundamentally quantum particle at all temperatures at which the metal exists as a solid. Indeed, according to the Heisenberg uncertainty principle, the momentum $\bar{p}$ of an electron is of order $\hbar/d$, where d is the lattice constant (it is the length characterizing the region of motion of the electron). The quantity d determines the spread in the position of the electron. The characteristic energy corresponding to this momentum (the so-called *Fermi energy*) is of order

$$\varepsilon \sim p^2/m \sim \hbar^2/(md^2); \quad \varepsilon \sim 5 \div 10 \text{ eV}, \tag{5.1}$$

and therefore $\varepsilon \gg T$ at all temperatures for which the metal exists as a solid. Hence the Dulong and Petit law $\varepsilon \sim T$ cannot be correct for the electrons in a metal: it contradicts the Heisenberg uncertainty principle. Therefore the electron cannot be described as a classical particle even from the qualitative point of view.

The Fermi energy (5.1) is intimately connected with the Pauli exclusion principle for the electrons: according to this principle not more than one electron can exist in each quantum state. For example, all of the electrons cannot have an energy corresponding to the lowest energy level. As the energy levels are successively filled with electrons (starting from the lowest level), the electrons occupy a sphere in momentum space, called the Fermi sphere. The momentum at the surface of the Fermi sphere corresponds to the energy (5.1). Therefore (5.1) can be regarded not only as the typical energy of an electron in the metal, but also as the maximum energy of the electrons in a metal in equilibrium.

It also follows from the Pauli exclusion principle that in any interaction only electrons whose energies are close to the Fermi energy are involved. These electrons can easily make transitions into the free states outside the Fermi sphere as a result of the interactions. This is much more difficult for the electrons lying deep within the Fermi sphere, since the neighboring states are occupied by other electrons and a large energy barrier must be overcome in order to make a transition into a free state. At a temperature T we can assume that only electrons whose energies differ from the Fermi energy by a quantity of order T can participate in interactions.

Using these ideas, we first estimate the electronic part of the heat capacity of a metal. Let $\varepsilon(T)$ be the energy of the electron at temperature T and $c = d\varepsilon(T)/dT$ be the heat capacity of the electron gas per conduction electron. We estimate the quantity c. In the zeroth approximation $\varepsilon(T)$ does not depend on T and is determined by the estimate (5.1). In essence, (5.1) corresponds to the zero-order term of a series expansion of the energy of the electron in T at $T = 0$. The next term in the expansion of $\varepsilon(T)$ in a Taylor series in T is linear in T. However, it is easily shown that this term must be equal to zero so that the electron gas will be stable to a spontaneous change in temperature T from the value $T = 0$. The stability condition has the form $d\varepsilon(T)/dT = 0$ at $T = 0$. Therefore the expansion of $\varepsilon(T)$ in T begins with the quadratic term in T:

$$\varepsilon(T) = \varepsilon + bT^2. \tag{5.2}$$

We estimate the constant b in (5.2). We use the fact that the expansion (5.2) becomes meaningless when $T \sim \varepsilon$: in this case the first and second terms in (5.2) are of the same order of magnitude. Therefore $\varepsilon \sim b\varepsilon^2$ and $b \sim \varepsilon^{-1}$. Then according to (5.2), the heat capacity of the electron in a metal has the form

$$c = \frac{d\varepsilon(T)}{dT} \sim \frac{d}{dT}\left(\frac{T^2}{\varepsilon}\right) \sim \frac{T}{\varepsilon} \ll 1. \tag{5.3}$$

In this example it is instructive to note that from the mathematical point of view the derivative dy/dx cannot always be approximated as the ratio y/x. It behaves as y/x only when the region in which y is significant in the problem under consideration corresponds to the characteristic region of variation in the independent variable x. In the case of (5.2) the significant values of y are large in comparison with the significant values of x and the estimate of the derivative, as seen from (5.3), is completely different.

It is evident from (5.3) that the heat capacity of the electron gas per unit volume is of order

$$C \sim \frac{T}{\varepsilon} n, \tag{5.4}$$

where n is the concentration of conduction electrons in the metal. The quantity (5.4) is small in comparison with the ionic part of the heat capacity of the metal, which, according to the law of Dulong and Petit, is of order n when the temperature is not too low. Because of the quantum nature of the electrons, the conduction electrons do not contribute significantly to the heat capacity of a metal. Here, we conclude our qualitative discussion of the equilibrium properties of metals and turn now to the basic kinetic characteristics.

5.1. Relation between the electrical and thermal conductivities of metals

As we will see, the thermal conductivity of a metal is due mainly to the conduction electrons, rather than to sound waves, as in a dielectric. Assuming this, we find a relation between the electronic part of the thermal conductivity λ of a metal and its electrical conductivity. Because of this relationship, we will consider below only the different effects influencing the thermal conductivity of the metal. Knowing the thermal conductivity λ, we may then easily estimate the electrical conductivity σ_e.

The flux density of thermal energy q transferred in response to a given (small) temperature gradient dT/dx is, according to (1.31):

$$q=-\lambda\frac{dT}{dx}\sim\varepsilon i. \tag{5.5}$$

Here, i is the flux density of electrons and ε is the energy of a single electron (5.1). The electric current density is then $j = -ei$. On the other hand, in a weak electric field E we have $j = \sigma_e E$ for the current density, according to Ohm's law. Here σ_e is the electrical conductivity of the metal.

The quantities E and dT/dx can be related to one another by noting that the energy acquired by an electron over a distance dx is equal to $d\varepsilon = eE\,dx$. On the other hand, this change of energy $d\varepsilon$ occurs because of the change in temperature dT. Therefore we can write

$$d\varepsilon=\frac{d\varepsilon}{dT}dT=c\,dT, \tag{5.6}$$

where c is the heat capacity of the electron gas per electron. Substituting (5.3) into (5.6), we obtain

$$d\varepsilon\sim\frac{T}{\varepsilon}dT. \tag{5.7}$$

Setting this expression equal to the quantity $eE\,dx$, we find the required relation between E and dT/dx:

$$E\sim\frac{T}{e\varepsilon}\frac{dT}{dx}. \tag{5.8}$$

Substituting (5.8) into (5.5), we find

$$\lambda\frac{dT}{dx}\sim\frac{\varepsilon\,\sigma_e}{e}\,\frac{T}{e\varepsilon}\frac{dT}{dx}, \tag{5.9}$$

and hence, canceling out the arbitrary temperature gradient dT/dx, we obtain a relation between the thermal conductivity λ (more exactly, its electronic part) and the electrical conductivity σ_e of the metal:

$$\lambda\sim\frac{T}{e^2}\sigma_e. \tag{5.10}$$

This relation is called the **Wiedmann–Franz law.**

We note that in the derivation of (5.10) the energy of the electron ε canceled out [this is obvious from (5.9)]. This implies that (5.10) has the same form if the conduction electrons are treated as classical particles with energies of the order of the temperature.

Problem 1. Show that the ratio $\lambda/(\sigma_e T)$ is of order 2×10^{-8} W Ω/K^2 for metals.

Problem 2. Show that for the Wiedmann–Franz law to be applicable the change in the energy of an electron in a collision with a thermal phonon or impurity must be small in comparison with T.

5.2. Residual resistance and impurity thermal conductivity

As noted above in the introduction to this chapter, an ideal lattice does not offer any resistance to a current of conduction electrons because of the coherence of the scattered electron waves from the ions of an ideal lattice. The coherence of the scattering can be destroyed for different reasons. An example is thermal motion of the ions about the lattice sites of the crystal. Because of its random nature, it destroys the coherence of the scattering and causes resistance, which is called *thermal resistance*. This type of resistance will be discussed in the next section. At low temperatures the thermal motion decreases and hence the thermal resistance also decreases. Here, another mechanism becomes important: the resistance caused by the scattering of electrons by impurities in the crystal lattice. The impurity centers are distributed randomly and therefore lead to incoherent scattering of the de Broglie electron waves. This incoherence is still present even in the case of an ordered arrangement of impurity centers, since their concentration is small and so the distance between impurity centers is large in comparison with the de Broglie wavelength $\lambda = \hbar/\bar{p} \sim d$ (d is the lattice constant of the crystal).

The mean free path of a conduction electron between collisions with impurity centers can be estimated with the help of the usual formula (1.1) for a gas; indeed, the concentration of impurity centers n_1 is assumed to be small in comparison with the concentration n of the principal ions making up the lattice, and therefore we can assume that the impurity centers form a rarefied gas. We therefore obtain

$$l \sim 1/(n_1 \sigma_{t1}). \tag{5.11}$$

Here, σ_{t1} is the cross section for momentum transfer in the scattering of an electron by the impurity atoms (or ions). In the case when the impurity centers are neutral atoms we have $\sigma_{t1} \sim d^2$, where d is the characteristic linear dimension of an impurity atom (or molecule). If the impurity is made up of ions, then

$$\sigma_{t1} \sim \left(\frac{e^2}{\varepsilon_0 \varepsilon}\right)^2 \tag{5.12}$$

according to (3.64). Here, we have substituted the energy of the electron ε in place of μv^2 and have neglected the Coulomb logarithm.

Knowing the mean free path l, we can estimate the electrical conductivity of the electron gas. It is simplest to start from (2.4):

$$\sigma_e \sim \frac{ne^2 l}{\bar{p}}. \tag{5.13}$$

Here, n is the concentration of electrons and $\bar{p}$ is the momentum of an electron. Setting $\bar{p} \sim \hbar/d$ on the basis of the Heisenberg uncertainty principle, we obtain the impurity electrical conductivity:

$$\sigma_e \sim \frac{n}{n_1} \frac{e^2 d}{\hbar \sigma_{t1}}. \tag{5.14}$$

The reciprocal of this quantity is the impurity electrical resistivity. It is also called the *residual resistivity*, since it remains even at low temperatures, when the thermal resistivity vanishes.

Because of differences in preparation, different samples of a given metal can have different concentrations of impurities and therefore, in contrast to the thermal resistivity, the impurity resistivity can be different in different samples.

Estimating the quantity d as the Bohr radius $\hbar^2 \varepsilon_0/(me^2)$, we obtain from (5.14) the following estimate for the impurity resistivity $\rho = \sigma_e^{-1}$:

$$\rho \sim \frac{n_1}{n} \frac{m\sigma_{t1}}{\hbar \varepsilon_0}. \tag{5.15}$$

Setting $\sigma_{t1} \sim d^2$ [this estimate is essentially correct also for ions, if we substitute (5.1) into (5.12)], we obtain a final expression for the impurity resistivity from (5.15):

$$\rho \sim \frac{n_1}{n} \frac{\hbar}{\varepsilon \varepsilon_0}, \tag{5.16}$$

where ε is the energy of the electron (5.1).

We now consider the impurity thermal conductivity λ of a metal; it is found most simply by using the Wiedmann–Franz law (5.10). Substituting (5.16) into (5.10), we find

$$\lambda \sim \frac{T}{e^2} \frac{n \varepsilon \varepsilon_0}{n_1 \hbar} \sim \frac{Tn}{\hbar n_1 d}. \tag{5.17}$$

We see that the impurity resistivity and conductivity do not depend on temperature, while the impurity thermal conductivity is proportional to T.

As noted in Problem 2 to the preceding section, for the Wiedmann–Franz law to be applicable, the scattering of the electron by an impurity center must be elastic. This is because of the large mass of an impurity center in comparison with the mass of the electron. In addition, the energy of the first excited state of the impurity center must be large in comparison with T.

Problem 1. Show that when there are two physically different mechanisms of electron collisions (for example, scattering by two types of impurities) the resultant electrical resistivity is larger than or equal to the sum of the resistivities due to each of the mechanisms individually. The relation $\rho = \rho_1 + \rho_2$ holds when the presence of one mechanism does not affect the other and is called *Matthiessen's rule.*

Problem 2. Prove that the uncertainty in the energy of a conduction electron is much smaller than the quantity $\hbar/\tau$, where τ is the mean free time of the electron as determined by collisions with impurities.

5.3. Kinetics of the interaction of conduction electrons with thermal vibrations of the ions of the crystal lattice

If the metal is sufficiently pure, the resistivity at room temperature is determined mainly by the interaction between the conduction electrons of the metal with thermal vibrations of the ions about the lattice points of the crystal, and the effect of impurities is only secondary. As in a dielectric, these vibrations are sound waves induced by the thermal motion of the ions; we discuss first the nature of these vibrations in a metal.

In a dielectric the formula (4.24) gave an estimate of the characteristic speed of sound. The quantity $e^2/(d\varepsilon_0)$ in this formula was the energy of the Coulomb interaction between neighboring ions (or atoms) on the lattice of the dielectric crystal. In a metal the ions interact with neighboring ions and also with the conduction electrons, and the interaction energy in both cases is of order $e^2/(d\varepsilon_0)$. It can also be written as ε [the energy of a conduction electron (5.1)], since the quantity d is of the order of the Bohr radius $\varepsilon_0\hbar^2/(me^2)$. Hence we conclude that the speed of sound in a metal is of the same order of magnitude as in a dielectric:

$$c_s \sim \sqrt{\frac{\varepsilon}{M}} \sim \sqrt{\frac{m}{M}}\, v \ll v, \tag{5.18}$$

and c_s is small in comparison with the thermal velocity v of the conduction electrons.

The condition for classical acoustic vibrations has the form of (4.6). It is evident from (4.6) that in a metal the Debye temperature is the same order of magnitude as in a dielectric. In the case of a metal it can be written in the form

$$\Theta = \hbar\sqrt{ne^2/(M\varepsilon_0)} \sim \sqrt{\frac{m}{M}}\,\varepsilon \ll \varepsilon. \tag{5.19}$$

We see that the Debye temperature Θ is small in comparison with the energy ε of a conduction electron.

In the temperature region $\Theta < T \ll \varepsilon$ we can treat the lattice vibrations as classical, yet at the same time the motion of the electrons is fundamentally quantum mechanical, since the energy of a conduction electron ε is determined by (5.1) and does not depend on temperature.

We determine the mean free path of the electrons due to scattering by the thermal vibrations of the ions. The cross section for such a collision is of order $\sigma \sim \xi_T^2$, where the amplitude ξ_T of the thermal vibration is given by (4.4). Hence from the formula (1.1) for a gas, we obtain the following estimate for the mean free path of an electron:

$$l \sim \frac{1}{n\xi_T^2} \sim \frac{e^2 d^3}{Td^3\varepsilon_0} \sim \frac{e^2}{T\varepsilon_0} \sim \frac{\varepsilon}{T}\, d \gg d. \tag{5.20}$$

Because $\varepsilon \gg T$, the mean free path is large in comparison with the lattice constant of the crystal. This inequality also determines the condition under which the "phonon gas" can be treated as an ideal gas.

Substituting (5.20) into the general expression (5.13) for the electric conductivity, we find

$$\sigma_e \sim \frac{ne^2 \varepsilon d}{Tp} \sim \frac{ne^2 \hbar}{mT}. \tag{5.21}$$

Therefore *the component of the resistivity of a metal due to the scattering of electrons by thermal vibrations of the lattice is proportional to the temperature.* The formal condition of applicability of (5.21) is $T \gg \Theta$. However, it turns out that this formula is quite accurate down to temperatures $T \gtrsim 0.1\Theta$.

Knowing the resistivity, we can use the Wiedmann–Franz law to estimate the component of the thermal conductivity of a metal due to scattering of electrons by thermal vibrations of the lattice ions. From (5.21) and (5.10), we obtain ($\Theta > T$)

$$\lambda \sim \frac{T}{e^2} \frac{ne^2 \hbar}{mT} \sim \frac{n\hbar}{m}. \tag{5.22}$$

We see that, in contrast to the impurity thermal conductivity (5.17), this quantity does not depend on temperature.

In addition to heat transport by the electrons, in a metal heat can also be transported by acoustic vibrations, as in a dielectric. The corresponding thermal conductivity is given by (4.32). We denote it by λ_1 and rewrite it in terms of the energy of an electron ε:

$$\lambda_1 \sim ne^2 \sqrt{\varepsilon}/(T\sqrt{M}\varepsilon_0). \tag{5.23}$$

Letting λ_2 be the thermal conductivity (5.22), we obtain for the ratio of (5.22) and (5.23)

$$\frac{\lambda_2}{\lambda_1} \sim \frac{\varepsilon_0 \hbar T \sqrt{M}}{me^2 \sqrt{\varepsilon}} \sim \frac{T}{\varepsilon} \sqrt{\frac{M}{m}} > 1. \tag{5.24}$$

Hence we conclude that *at room temperature the thermal conductivity in pure metals is determined mainly by energy transport by electrons, not phonons.* However, the difference between the electron and phonon parts of the thermal conductivity is not large. Metals do conduct heat better than dielectrics; but in alloys the thermal conductivities are approximately the same order of magnitude as in dielectrics.

We estimate the drift velocity of electrons in a metal under an applied electric field E. The flux density of electrons is

$$i = nV = -\frac{1}{e} j = -\frac{\sigma_e E}{e}. \tag{5.25}$$

Here, V is the drift velocity of the electrons and it is defined by the above equation:

$$V = -\frac{\sigma_e E}{en}. \tag{5.26}$$

Substituting (5.21) for the conductivity of a metal into (5.26), we obtain

$$V \sim \frac{e\hbar}{mT} E. \tag{5.27}$$

The quantity $E_a \sim e^2/(d^2\varepsilon_0)$ ($E_a \sim 10^{11}$ V/m) is the typical electric-field strength in an elementary cell of the lattice. In terms of this quantity we can rewrite (5.27) in the form

$$V \sim \frac{E}{E_a} \frac{\varepsilon}{T} v \ll v. \tag{5.28}$$

For example, for a field of strength $E \sim 1$ V/m we obtain $V \sim 1$ mm/s.

Problem 1. Obtain (5.22) for the thermal conductivity of a metal using (1.34) for the thermal conductivity of a gas, (5.4) for the electronic part of the heat capacity C, and (5.20) for the mean free path, without using the Wiedmann–Franz law.

Problem 2. Obtain the law of proportionality of the resistivity to the temperature from the fact that when $T > \Theta$ the number of phonons per vibration is of order $T/(\hbar\omega_0)$, where ω_0 is the characteristic frequency of vibration (4.3).

Problem 3. Show that at high temperatures $T > \Theta$ equilibrium in the phonon gas is established much more rapidly because of phonon-phonon collisions than because of phonon-electron collisions.

5.4. Electron–electron collisions

Up to now we have neglected the possibility of collisions between conduction electrons. As a result of the Pauli exclusion principle only electrons whose energies are near the Fermi energy (5.1) can participate in collisions. The number of such electrons is very small in comparison with the total number of electrons and therefore electron-electron collisions are rare. The fraction of electrons which can scatter with each other depends on the temperature T. At $T = 0$ scattering is not possible in general. When $T \neq 0$ this fraction can be estimated in the following way: electron-electron scattering is the only contribution to the electronic part of the heat capacity of a metal. It follows from (5.3) that this fraction of electrons is of order T/ε. Hence the probability of collision of two electrons is of order $(T/\varepsilon)^2$. The collision frequency ν is proportional to this probability, i.e.,

$$\nu = b(T/\varepsilon)^2. \tag{5.29}$$

The proportionality constant in this expression is found using the fact that when $T \sim \varepsilon$ we must have $\nu \sim v/d$, since then the electrons collide over a distance of the order of the lattice constant d. Hence we obtain the following estimate for the collision frequency:

$$\nu \sim \frac{v}{d} \left(\frac{T}{\varepsilon}\right)^2 \sim \left(\frac{T}{\varepsilon}\right)^2. \tag{5.30}$$

We can now estimate the mean free path of the electrons as determined by collisions with other electrons in the metal:

$$l \sim \frac{v}{\nu} \sim \frac{v\hbar\varepsilon}{T^2} \sim \left(\frac{\varepsilon}{T}\right)^2 d. \tag{5.31}$$

We compare this quantity (denote it as l_{ee}) with the mean free path of the electrons as determined by collisions with acoustic lattice vibrations (5.20) (denote this quantity as l_{ei}):

$$l_{ee}/l_{ei} \sim \varepsilon/T \gg 1. \tag{5.32}$$

Hence at *room temperature electron-phonon collisions dominate.*

Electron-electron collisions become important in pure metals at very low temperatures, when the mean free path l_{ei} as determined by collisions with phonons is no longer given by (5.20), but is much larger (see Sec. 5.5).

Problem 1. Show that at room temperature the time τ_{ee} between electron-electron collisions in a metal is of order 10^{-10} s.

Problem 2. Show that in metals electron-electron interactions are screened over distances of the order of the lattice constant of the crystal.

Problem 3. Obtain the following estimate for the component of the conductivity of a metal due to electron-electron collisions:

$$\sigma_{ee} \sim \frac{ne^2\hbar}{m\varepsilon}\left(\frac{\varepsilon}{T}\right)^2.$$

Problem 4. Show that the contribution to the thermal conductivity of a metal due to electron-electron collisions is inversely proportional to the temperature.

5.5. Thermal and electrical conductivities of metals at low temperatures

In this section we consider the properties of metals at low temperatures $T \ll \Theta$. In this case only phonons with energies $\hbar\omega \sim T$ interact with electrons, and the change in energy of an electron in a collision with a phonon is of order $\Delta\varepsilon \sim T$. The number of phonons with energies $\hbar\omega \gg T$ is exponentially small, according to Boltzmann statistics. Here, we are concerned with long-wavelength acoustic phonons, since the energy of the optical phonons $\hbar\omega \gg T$.

The probability w of collision of an electron with a phonon is proportional to the number of phonons $d\mathbf{k}$ per unit volume ($\mathbf{k}$ is the wave vector of the phonon). Because for an acoustic phonon $\omega = c_s k$, where c_s is the speed of sound, it follows that the quantity w is proportional to ω^3, i.e., T^3. Therefore the number ν of collisions of an electron with phonons per unit time is proportional to the cube of the temperature. The proportionality constant is found from the condition that at $T \sim \Theta$ the frequency ν should be the same order of magnitude as obtained from the high-temperature limit for ν as T is decreased down to Θ. This limit is obtained from (5.20) for the mean free path l:

$$\nu \sim \frac{v}{l} \sim \frac{v\varepsilon_0 T}{e^2} \sim \frac{vT}{\varepsilon d} \sim \frac{T}{\hbar} \quad (T \gg \Theta). \tag{5.33}$$

When $T \sim \Theta$ we obtain $\nu \sim \Theta/\hbar$.

Having determined the proportionality constant between ν and T, we find

$$\nu \sim \frac{\Theta}{\hbar}\left(\frac{T}{\Theta}\right)^3 \quad (T \ll \Theta). \tag{5.34}$$

We can now obtain the mean free path of electrons as determined by collisions with acoustic phonons for $T \ll \Theta$:

$$l = l_{ei} \sim \frac{v}{\nu} \sim \frac{\hbar v \Theta^2}{T^3}. \tag{5.35}$$

These results can now be substituted into (1.34), $\lambda \sim Cvl$ for the thermal conductivity. Here, C is the electronic part of the heat capacity per unit volume of the metal and is given by (5.4). Substituting (5.4) and (5.35) into (1.34), we find

$$\lambda \sim \frac{T}{\varepsilon} nv \frac{\hbar v \Theta^2}{T^3} \sim \frac{n\hbar}{m}\left(\frac{\Theta}{T}\right)^2. \tag{5.36}$$

We see that λ is inversely proportional to the square of the temperature in the region $T \ll \Theta$; but when $T \gg \Theta$ it follows from (5.22) that it is independent of temperature. Hence *the thermal conductivity of a metal increases as the temperature is lowered.*

The contributions to the thermal conductivity from electron-electron and electron-phonon collisions at low temperatures can be compared by comparing the corresponding mean free paths (5.31) and (5.35):

$$\frac{l_{ee}}{l_{ei}} \sim \frac{\varepsilon^2 d}{T^2} \frac{T^3}{\hbar v \Theta^2} \sim \frac{T}{\Theta} \frac{\varepsilon}{\Theta}. \tag{5.37}$$

We have $T \ll \Theta$, but $\varepsilon \gg \Theta$ and so there is a competition between the two factors in (5.37). The electron-phonon mechanism dominates when the temperature is not too low. The electron-electron mechanism becomes important at very low temperatures in a metal of extremely high purity.

We turn now to the conductivity of pure metals at low temperatures. As we have seen in Sec. 1.3, the thermal conductivity is associated with heat transfer by means of collisions. In each collision between an electron and a phonon the energy of the electron changes by a quantity of the order of the energy of the phonon T (an example is absorption of a phonon with an energy of order T). Therefore the time between two successive collisions (the mean free time) is also the energy relaxation time. This justifies the use of the mean free time $\tau \sim l/v$ as the relaxation time for heat conduction.

However, the electrical conductivity is associated with momentum transfer, rather than energy transfer. We consider how momentum is transferred in a collision between an electron and a phonon at low temperatures. The momentum of the phonon is of order $\hbar k \sim T/c_s$. By conservation of momentum, the momentum of the electron changes by the same amount $\overline{\Delta p} \sim T/c_s$. On the other hand, the change in momentum Δp and energy $\Delta\varepsilon$ of the electron are related by the equation

$$\Delta\varepsilon \sim p\Delta p \cos\theta/m \sim vT\cos\theta/c_s. \tag{5.38}$$

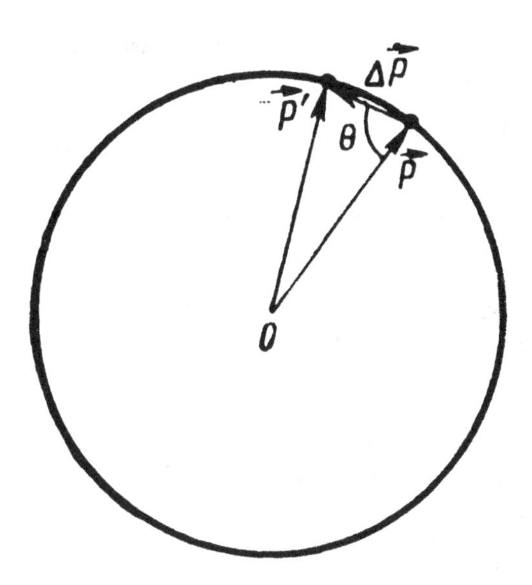

FIG. 4. Diffusion of an electron on the Fermi surface due to collisions with the phonons of the crystal lattice.

Here, θ is the angle between $\Delta\bar{p}$ and $\bar{p}$. Since $\Delta\varepsilon \sim T$, the angle θ is close to 90° and therefore in each collision the momentum of the electron changes in direction by a small angle on the Fermi sphere (Fig. 4).

Therefore at low temperatures electron-phonon collisions lead to two-dimensional diffusion of electrons on the Fermi surface in momentum space. In each collision the square of the momentum changes by a small quantity $(\Delta\bar{p})^2 \sim (T/c_s)^2$. We want to determine the number of collisions N after which the momentum of the electron changes by a quantity of the order of the momentum $\bar{p}$ itself, i.e., the momentum vector of the electron changes in direction by a significant amount (by an angle of order 1 rad) as a result of its diffusion (and not from regular changes of direction by a small angle). This number N is given by the condition

$$N(T/c_s)^2 \sim p^2, \tag{5.39}$$

which is analogous to the conditions of Sec. 1.5 for momentum diffusion. We obtain from (5.39)

$$N \sim (pc_s/T)^2 \sim (\Theta/T)^2. \tag{5.40}$$

Therefore the momentum relaxation time $\tau_{\bar{p}}$ is a factor of N larger than the energy relaxation time $\tau_e \sim \nu^{-1}$, where ν is given by the estimate (5.34). We obtain

$$\tau_p \sim \frac{\hbar}{\Theta}\left(\frac{\Theta}{T}\right)^5. \tag{5.41}$$

We can now estimate the conductivity of a metal at low temperatures $T \ll \Theta$. We rewrite (5.13) in the form

$$\sigma_e \sim \frac{ne^2\tau_p}{m}. \tag{5.42}$$

Here, we have used the relation between the mean free path and the collision time $l \sim v\tau_{\bar{p}}$. Substituting (5.41) into (5.42), we finally obtain

$$\sigma_e \sim \frac{ne^2\hbar}{m\Theta}\left(\frac{\Theta}{T}\right)^5. \tag{5.43}$$

Hence *the electrical conductivity is inversely proportional to the fifth power of the temperature* **(Bloch's law).**

Note that when $T \sim \Theta$ the estimates (5.43) and (5.21) are the same order of magnitude, as one would expect. Similarly, the estimates (5.36) and (5.22) for the thermal conductivity coincide in the region $T \approx \Theta$.

Problem 1. Show that in a strong electric field the field-dependent correction to the conductivity of a metal is proportional to the square of the electric-field strength.

Problem 2. Assuming $T \ll \Theta$, obtain the estimate

$$\sigma_e \sim \exp[\hbar\omega_0/T]$$

for the component of the conductivity of a metal due to the interaction of electrons with optical vibrations of the crystal lattice at frequency ω_0.

Problem 3. Show that at low temperatures in semiconductors scattering of charge carriers (electrons or holes) by ionized lattice defects dominates, and the mean free time of the charge carriers is $\tau \sim T^{3/2}$. Show also that at high temperatures the dominant mechanism is scattering of charge carriers by acoustic phonons and $\tau \sim T^{-3/2}$.

Problem 4. Show that for semiconductors the determining factor in the dependence of the conductivity on temperature is not the mean free time of the charge carriers, but the concentration of charge carriers, which depends exponentially on temperature.

Chapter 6

Flow of an ideal fluid

By an ideal fluid we mean a fluid in which one can neglect all relaxation processes: viscosity, heat conduction, and so on. We first obtain the general equations satisfied by the velocity $\mathbf{v}$ of an ideal fluid. Because the compressibility of a fluid is small in most cases, we will assume that the fluid is incompressible. We obtain the condition of incompressibility in mathematical form.

We consider a cube of unit volume in the fluid. Because of the incompressibility of the fluid, the quantity of fluid flowing into the cube per unit time must be equal to the quantity of fluid flowing out per unit time. Suppose the velocity of the fluid is along the x axis. We write the velocity as v_x. The flux density of fluid passing through a side of the cube perpendicular to the x axis is equal to ρv_x. The difference between the flux densities through the opposing sides of the cube must be zero because of incompressibility, i.e., we have $d(\rho v_x)/dx = 0$. The density ρ of an incompressible fluid is constant and therefore it can be eliminated from the problem.

We generalize the above relation to the case when the fluid also moves along the y and z axes:

$$\partial v_x/\partial x + \partial v_y/\partial y + \partial v_z/\partial z = 0, \tag{6.1}$$

or $\operatorname{div} \mathbf{v} = 0$.

We consider now the equation of motion (Newton's second law) for a unit cube of fluid assuming that the velocity of the fluid is along x. The force acting on the cube is equal to the pressure difference on the opposing sides of the cube perpendicular to the x axis and is given by $\partial P/\partial x$. Therefore we obtain

$$\rho \frac{dv_x}{dt} = -\frac{\partial P}{\partial x}.$$

The generalization of this equation to the case of three-dimensional motion is obvious:

$$\rho \frac{d\mathbf{v}}{dt} = -\nabla P. \tag{6.2}$$

Applying the curl operation to (6.2), the right-hand side vanishes, since the curl of the gradient of any scalar function is equal to zero. The relation $d(\operatorname{curl} \mathbf{v})/dt = 0$ means that the quantity $\operatorname{curl} \mathbf{v}$ is constant along the streamlines. Because the velocity of the fluid is zero at infinity (or constant, depending on the choice of coordinate system), it follows that curl $\mathbf{v}=0$. This conclusion is not correct near the surface of a body moving in the fluid, where there are discontinuities in the streamlines. It is also incorrect in other cases where there are discontinuities in the streamlines, which will be discussed later.

The condition curl $\mathbf{v}=0$ means that we can introduce the so-called *velocity potential* with the help of the relation $\mathbf{v}=\nabla\varphi$. Then the condition curl $\mathbf{v}=0$ is satisfied identically and the flow is called *potential flow*. The equation expressing the incompressibility of the fluid, (6.1), then takes the form

$$\text{div grad}\,\varphi=0, \text{ or } \Delta\varphi=0. \tag{6.3}$$

The quantity

$$\Delta=\frac{\partial^2}{\partial x^2}+\frac{\partial^2}{\partial y^2}+\frac{\partial^2}{\partial z^2}$$

is called the *Laplacian*. Applying the gradient operator ∇ to (6.3) and using the easily verified relation $\nabla\Delta = \Delta\nabla$, we find that the velocity $\mathbf{v}$ satisfies the same equation as φ (called **Laplace's equation**):

$$\Delta\mathbf{v}=0. \tag{6.4}$$

6.1. Flow of a fluid around a body

We estimate the velocity field of an ideal incompressible fluid when a body of characteristic linear dimension R moves in it with a constant velocity u. We consider the qualitative solution of (6.4). It is easy to show that the function $1/r$ is a spherically symmetric solution of Laplace's equation. Therefore suppose we construct the qualitative solution in the form $\mathbf{v}\sim(R/r)\mathbf{u}$. Here, we have used the fact that we must have $v\sim u$ at $r\sim R$. However, it can easily be verified that in this case the total quantity of fluid carried along by the body is infinite. Indeed, the flux density ρv is of order $1/r$; integrating it over the area of a cross section $dS\sim r dr$ perpendicular to the direction of the flow, we obtain a divergent integral. Hence this solution is no good.

The function $\nabla(1/r)$ is also a solution of Laplace's equation since, as noted above, $\Delta\nabla=\nabla\Delta$. However, if $v\sim r^{-2}$ the total quantity of fluid carried along by the body is equal to a logarithmically divergent integral. Therefore this solution also is not suitable.

Applying the gradient operation one more time, we obtain the velocity in the form

$$\mathbf{v}\sim(\mathbf{u}\nabla)\nabla(R^3/r), \tag{6.5}$$

which satisfies all of the necessary physical requirements: the quantity (6.5) satisfies Laplace's equation (6.4) and leads to a finite quantity of fluid carried along by the body.

In the limit $r\gg R$ we obtain from (6.5) the following estimates for the velocity and velocity potential:

$$v\sim(R/r)^3u, \quad \varphi\sim(R/r)^2uR. \tag{6.6}$$

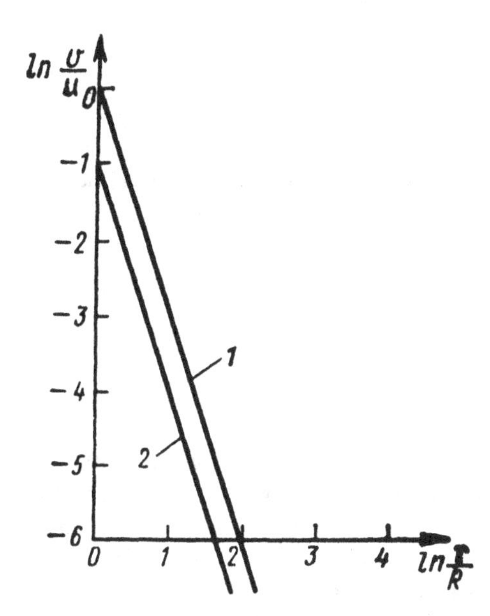

FIG. 5. Velocity v of a fluid induced by a sphere of radius R moving in it with velocity u as a function of the distance r from the center of the sphere: (1) along the velocity vector of the sphere; (2) perpendicular to the velocity of the sphere; the slope of the lines is equal to − 3.

The proportionality constant in (6.6) depends on the angle between the position vector $\mathbf{r}$ of the observation point and the velocity of the body, as well as on the shape of the body. In the case of a sphere of radius R the angular dependence is exactly given by (6.5), since this combination of the vectors $\mathbf{u}$ and ∇ is the only one possible. The exact solution for the velocity $\mathbf{v}$ in the case of a sphere is shown in Fig. 5. Note that at large distances r from the moving body the velocity of the fluid falls off rapidly to zero.

Next, we estimate the velocity field of potential flow around an infinitely long cylinder of radius R (or any long body with a characteristic linear transverse dimension R). The cylinder moves with a constant velocity u perpendicular to its axis in an ideal incompressible fluid.

The Laplacian in Laplace's equation (6.4) is now a two-dimensional operator. It is easy to verify that an angle-independent solution of this equation is $v \sim \ln r$. In this case the quantity of fluid carried along by the cylinder per unit length of the cylinder must be finite. Multiplying $\ln r$ by the cross-sectional area $dS \sim dr$, we obtain a divergent integral. Furthermore, a solution of the form $v \approx \nabla (\ln r)$ leads to a logarithmically divergent integral. The next order of differentiation with respect to r gives a solution satisfying the requirement that the total quantity of fluid carried along by the body per unit length must be finite. Hence, in analogy to (6.5) we obtain

$$\mathbf{v} \sim (\mathbf{u}\nabla)\nabla(R^2 \ln r). \tag{6.7}$$

At large distances $r \gg R$ we obtain approximate expressions for the velocity and velocity potential:

$$v \sim (R/r)^2 u, \quad \varphi \sim uR^2/r. \tag{6.8}$$

We see that the velocity falls off more slowly with r than in the preceding example.

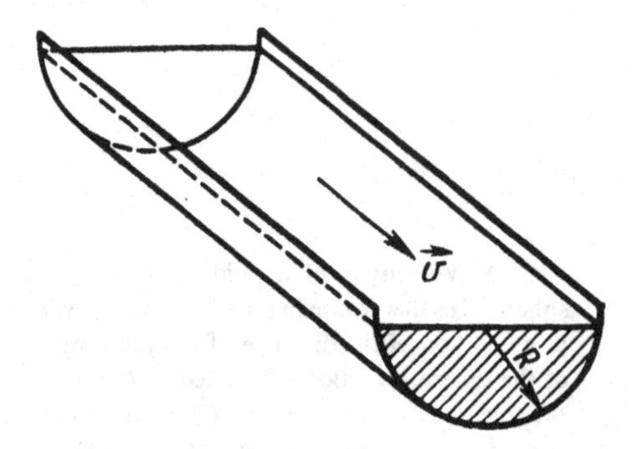

FIG. 6. Flow of a fluid from an inclined trough of radius R under gravity.

The higher-order derivatives with respect to r will also be solutions of Laplace's equation; however, they fall off at large distances as higher powers of $1/r$ and hence they are small in comparison with (6.5) and (6.7). Therefore they do not change the estimates (6.6) and (6.8).

We note that the results for these problems cannot be obtained purely by dimensional analysis, since the ratio R/r is a dimensionless quantity.

In other problems dimensional considerations may be sufficient for a qualitative solution. For example, suppose that an ideal incompressible fluid occupies all of space and the fluid is removed from a cavity with linear dimension R. We estimate the time required for the fluid to again fill up the cavity. The pressure P of the fluid is given. The required time τ is determined by the pressure P of the fluid, its density ρ, and obviously the dimensions of the cavity R. Only one combination of these quantities has the units of time. Since the dimensions of pressure are kg/(m s^2), we find that $[P/\rho] = (\text{m/s})^2$ (here and below the square brackets $[f]$ denote the dimensions of the quantity f). Therefore $[(\rho/P)^{1/2}] = \text{s/m}$ and finally

$$\tau \sim R(\rho/P)^{1/2}. \tag{6.9}$$

In particular, for a sphere of radius R a numerical calculation shows that the proportionality constant in (6.9) is equal to 0.91.

We see from (6.9) that the velocity of collapse of the cavity is of order $(P/\rho)^{1/2}$ and is independent of the dimensions of the cavity R.

Problem 1. A liquid flows from an inclined trough of linear dimension R (Fig. 6) under gravity. Show that the flow rate is proportional to $R^{5/2}$.

Problem 2. Show that when a body of mass m falls vertically with velocity v into a fluid, the maximum force of impact is proportional to v^2 and to $m^{2/3}$.

Problem 3. A body with density ρ_1 slightly larger than the density of the fluid in which it is immersed, begins to sink. Show that the initial acceleration of the body is of order $(\rho_1 - \rho)g/\rho$, where g is the acceleration of gravity.

Problem 4. A liquid rotates in a bucket of radius R with a constant angular velocity ω. Obtain the estimate $h \sim \omega^2 R^2/g$ for the depth h of the resulting vortex.

6.2. Gravity waves on the surface of a liquid

The propagation of surface gravity waves in a liquid is determined by the force of gravity. Therefore the velocity c_g of the wave should be determined by the acceleration of gravity g. In addition, in general it also depends on the wavelength λ. We estimate the velocity of gravity waves with a given wavelength λ on the surface of a deep liquid in which we can neglect the effects of the bottom.

We first discuss the dependence of c_g on the amplitude a of the wave (which characterizes the intensity of the wave process). If $a \ll \lambda$ the velocity c_g can be expanded in a Taylor series in the small dimensionless parameter $(a/\lambda) \ll 1$. It is evident from physical considerations that the expansion contains only even powers of the parameter, since the vibration amplitude a can be positive or negative. The zero-order term of the expansion, in general, does not depend on a. Therefore we can estimate c_g by constructing a quantity with the dimensions of velocity from the combination of λ and the acceleration of gravity g. Since $[g] = \mathrm{m/s^2}$, we obtain for the wave velocity c_g

$$c_g \sim (g\lambda)^{1/2}. \tag{6.10}$$

The numerical factor in this dependence is $(8\pi)^{-1/2}$. From the above discussion it is evident that (6.10) determines the velocity of surface gravity waves of small amplitude.

According to (6.10), the velocity c_g depends on the wavelength λ and hence the waves have *dispersion*.

In obtaining (6.10) we neglected the pressure of the liquid, since at the surface it approaches zero. In addition, we assumed that the depth h of the liquid is large in comparison with the depth over which the wave disturbance extends. The latter depth is of the order of the wavelength λ. Here, it should be emphasized that the wave amplitude a, which also characterizes the path length traveled by a fixed particle of liquid over a period of the wave, is small in comparison with the depth over which the wave disturbance extends and does not in any way determine this depth. To see this, note that we consider the wave as a linear process in which the typical velocity of a particle in the wave and its position are proportional to a. Therefore the amplitude a cannot appear in expressions determining the damping of the velocity of the liquid with depth. Hence the damping with depth is determined only by the wavelength λ.

Therefore (6.10) holds when

$$h \gg \lambda \gg a. \tag{6.11}$$

From (6.10) for the wave velocity c_g we can estimate the dependence of the wave frequency ω on its wavelength λ. Since $\omega \sim c_g/\lambda$ (here we do not distinguish between the phase and group velocities of the wave, since we are not interested in numerical factors), we find from (6.10)

$$\omega \sim (g/\lambda)^{1/2}. \tag{6.12}$$

The quantitative solution of the problem must be based on (6.1) and (6.2). On the left-hand side of (6.2) the velocity $\mathbf{v}$ depends on the time t both explicitly and also implicitly through the position $\mathbf{r}$ of a fixed particle of liquid, which may also vary with time t. Therefore (6.2) can be rewritten in the form

$$\rho\left[\frac{\partial \mathbf{v}}{\partial t}+(\mathbf{v}\nabla)\mathbf{v}\right]=-\nabla P+\rho\mathbf{g}. \tag{6.13}$$

On the right-hand side of (6.13) we have added the force of gravity acting on a unit volume of liquid.

The second term on the left-hand side of (6.13) is nonlinear in the velocity $\mathbf{v}$. In the linear approximation this term is neglected. We obtain the condition when it is possible to do this. The velocity v of a particle of liquid can be estimated as $a\omega$. Therefore the condition $(\mathbf{v}\cdot\nabla)\mathbf{v}\ll\partial\mathbf{v}/\partial t$ can be written qualitatively in the form $v^2/\lambda \ll \omega v$ or $a\omega\ll\lambda\omega$ (since the velocity varies over a linear distance of the order of a wavelength λ; see above). Hence when $a\ll\lambda$ the equations of motion for the wave process are linear. This is consistent with the above discussion.

We now estimate the velocity of gravity waves on the surface of a shallow liquid, whose depth satisfies the condition $h\ll\lambda$, which is opposite to the condition assumed in the preceding case. It corresponds to the long-wavelength limit. In place of (6.11) we now have the condition

$$a\ll h\ll\lambda, \tag{6.14}$$

where a is the amplitude (height) of the wave. In the limit of very long wavelength λ (or very low frequency ω) the dependence of the wave velocity c_g on frequency disappears. We note that an expansion of c_g in powers of ω contains only even powers of ω since the sign of the frequency ω is arbitrary and the wave velocity cannot depend upon it.

Hence we need to construct a quantity with the dimensions of velocity from the acceleration of gravity g and the depth of the liquid h. We find

$$c_g\sim(gh)^{1/2}. \tag{6.15}$$

We note that the numerical factor in this expression is equal to unity.

It follows from (6.15) that unlike the case of a deep liquid, in a shallow liquid gravity waves do not have dispersion, i.e., the velocity c_g of the wave does not depend on λ. Furthermore, we find for the frequency $\omega \sim c_g/\lambda$ of the wave

$$\omega\sim(gh)^{1/2}/\lambda. \tag{6.16}$$

For waves in a shallow liquid the vertical component of the velocity of liquid particles is obviously small in comparison with the horizontal component, unlike the case of a deep liquid, where the two velocities are of the same order of magnitude. The horizontal component v_x of the velocity varies only slightly over the entire depth of the liquid, i.e., it is practically independent of the vertical coordinate z. In contrast, an expansion of the vertical component of the velocity v_z in a Taylor series in z leads to the relation $v_z=$ const z, since we must have $v_z=0$ at the bottom $(z=0)$. Therefore the vertical component of the velocity at the surface of the liquid $z=h$ is $v_z=$ const h. The proportionality constant in this dependence is found using the fact that when $h\sim\lambda$ we must have $v_z\sim v_x$, because then the problem transforms into the preceding problem, where the vertical and horizontal components of the velocity are of the same order of magnitude. Therefore we obtain

$$v_z \sim \frac{h}{\lambda} v_x \ll v_x. \tag{6.17}$$

This relation can also be obtained from the incompressibility condition (6.1), since the velocity v_z varies along the z axis over the characteristic distance h, while v_x varies along the x axis over the characteristic distance λ, and therefore from (6.1) we have $v_z/z \sim v_x/x$.

Problem 1. An underwater explosion occurs at a depth h in a liquid. The energy released in the explosion is E and an oscillating gas bubble forms as a result of the explosion. Show that the period of oscillation is of order $(E/\rho)^{1/3}(gh)^{-5/6}$, where ρ is the density of the liquid.

Problem 2. An incompressible fluid sphere of radius R performs quadrupole oscillations as a result of the gravitational attraction between the particles of the fluid. Obtain the estimate

$$T \sim (R/g)^{1/2}$$

for the period of the oscillations. Here, g is the acceleration of gravity at the surface of the sphere.

Problem 3. The temperature T of air above a certain horizontal boundary in the atmosphere is higher than the temperature of the air below this boundary by the small quantity $\Delta T \ll T$. Show that gravity waves can propagate along the boundary between the two regions and the wave velocity c_g as a function of the wavelength λ is roughly

$$c_g \sim \left[\frac{\Delta T}{T} g\lambda\right]^{1/2}.$$

Chapter 7

Viscous gases and liquids

The dynamical viscosity η is the proportionality constant between the frictional force per unit area acting between neighboring fluid layers because of their different velocities, and the velocity gradient in the system:

$$F_x = -\eta \frac{dv_x}{dy} \tag{7.1}$$

[see (1.35)]. Here, F_x is the frictional force along the x axis and has units of pressure (force per unit area). Therefore $[F_x] = \mathrm{Pa} = \mathrm{kg/(m\,s^2)}$. It follows from (7.1) that the dimensions of the dynamical viscosity are $[\eta] = \mathrm{kg/(m\,s)}$. The viscosity was estimated for different cases in Sec. 1.4.

We generalize the equation of motion of a fluid (6.2) to the case when its viscosity must be taken into account. We again consider a unit volume of fluid in the shape of a cube. Suppose the fluid flows along the x axis. The viscous force (7.1) acts along the x axis on the sides of the cube perpendicular to the y axis, where it is assumed that the velocity of the fluid v_x varies along y. The net force is equal to the difference of the forces acting on the opposing sides of the cube (Fig. 7). According to (7.1), this difference is equal to $\eta d^2 v_x/dy^2$. Adding this term to the x component of the vector equation (6.2), we obtain

$$\rho \frac{dv_x}{dt} = -\frac{\partial P}{\partial x} + \eta \frac{\partial^2 v_x}{\partial y^2}. \tag{7.2}$$

This equation is simply Newton's second law for a fluid with the force of friction taken into account.

If the velocity varies v_x in all directions, and not just y, then instead of $\partial^2/\partial y^2$ in (7.2) we will have the Laplacian

$$\Delta = \frac{\partial^2}{\partial x^2} + \frac{\partial^2}{\partial y^2} + \frac{\partial^2}{\partial z^2}.$$

In addition, in the general case where the motion is not only along x, but also along the other axes, we must generalize (7.2) in the obvious way by replacing the scalar v_x by the vector $\mathbf{v}$. The resulting vector equation for a unit volume of viscous fluid is called the **Navier–Stokes equation**:

$$\rho \frac{d\mathbf{v}}{dt} = -\nabla P + \eta \Delta \mathbf{v}. \tag{7.3}$$

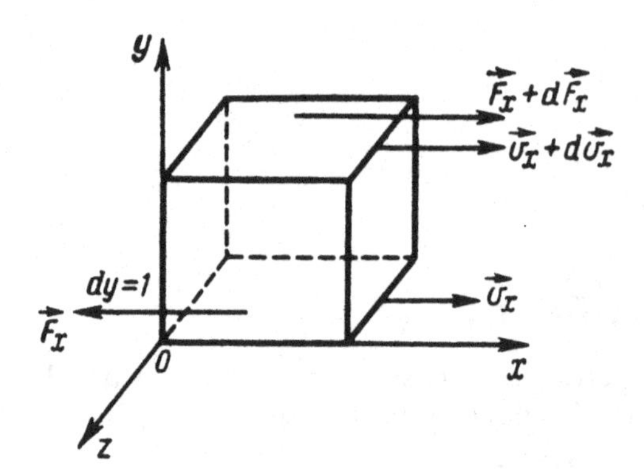

FIG. 7. Frictional force acting on a volume of fluid in the presence of a gradient in the velocity v_x in the y direction.

The left-hand side of this equation contains the total derivative with respect to time; the explicit form of this derivative was given in (6.13).

In the case of steady motion, when the velocity does not depend explicitly on time, (7.3) simplifies to

$$\rho(\mathbf{v}\nabla)\mathbf{v} = -\nabla P + \eta\Delta\mathbf{v}. \tag{7.4}$$

In this chapter we consider the case when the fluid velocities are sufficiently small. The left-hand side of (7.4) can be estimated as $\rho v^2/l$, where l is the typical scale of length of the flow field. Similarly, for the viscous term on the right-hand side of (7.4) we have the estimate $\eta v/l^2$. The ratio of the first quantity to the second is equal to vl/ν. Here, $\nu = \eta/\rho$ is the *kinematic viscosity*. The ratio $\mathrm{Re} = vl/\nu$ is called the *Reynolds number*. We conclude that when $\mathrm{Re} \ll 1$ the nonlinear term in the velocity on the left-hand side of (7.4) can be neglected. We then obtain

$$\nabla P = \eta\Delta\mathbf{v}. \tag{7.5}$$

Note that (7.4) is written in a coordinate system in which a body placed in the fluid is at rest, while at infinity there is steady stream of the fluid, which flows around the body. We can transform from this coordinate system to a coordinate system in which the fluid is at rest at infinity and the body moves in the fluid, by adding the constant velocity $\mathbf{u}$ of the body to all velocities in (7.4). Unlike (7.4), the simplified equation (7.5) does not change under this transformation.

7.1. Flow through pipes and pores

We estimate first the velocity of a viscous fluid flowing through a pipe of length l with characteristic linear transverse dimension R, under a small pressure difference δP. From physical considerations it is evident that the velocity of the fluid $v = v_x$ in the pipe (the x axis is taken along the axis of the pipe) is determined by the ratio $\delta P/l$, i.e., by the pressure gradient [this is also evident from (7.5)]. The pressure gradient has the dimensions $\mathrm{kg/(m^2\,s^2)}$. We construct a quantity with the dimensions of velocity from η, $\delta P/l$, and the transverse dimension of the pipe R:

$$v \sim R^2\,\delta P/(\eta l). \tag{7.6}$$

In agreement with (7.5), the density of the fluid ρ does not appear separately as a parameter of the problem. The result (7.6) holds when $\mathrm{Re} \ll 1$, i.e., when the velocity of the fluid is sufficiently small.

The expression (7.6) is the zero-order term of a Taylor-series expansion of the velocity v in the small quantity $\mathrm{Re} \ll 1$. When the Reynolds number is not negligibly small, (7.6) can be improved by taking into account the first-order term of the Taylor series

$$v \sim \frac{R^2\,\delta P}{\eta l}(1+C\,\mathrm{Re}) = \frac{R^2\,\delta P}{\eta l}\left(1+C\frac{\rho\,\delta P R^3}{\eta^2 l}\right), \tag{7.7}$$

where C is a numerical factor. We see from (7.7) that the second term can be neglected if the radius of the pipe R is sufficiently small. Flow described by (7.6) is called *Poiseuille flow.*

We estimate the quantity Q of fluid passing through the pipe per unit time (the flow rate). From the definition of this quantity it follows that $Q \sim \rho v R^2$. Substituting (7.6) into this relation, we obtain for the mass flow rate Q:

$$Q \sim \rho\,\delta P\,R^4/(\eta l). \tag{7.8}$$

From numerical calculations it can be shown that for a pipe of circular cross section of radius R the proportionality constant in (7.8) is equal to $\pi/8$, and then (7.8) is called *Poiseuille's law.*

We note that the value of the numerical factor in (7.8) is of order unity. This will normally be the case, since the hydrodynamic equation (7.4) does not contain very large or very small dimensionless parameters: the coefficients of the equation are of order unity. For this reason numerical factors differing significantly from unity cannot appear in the solutions to this equation.

However, suppose instead of a circular pipe we consider a pipe whose cross section is an equilateral triangle of side R. It follows from numerical calculations that the proportionality constant in (7.8) in this case is equal to $3^{1/2}/320 \ll 1$. Why is the numerical factor in this example so small, in apparent contradiction to the assertion of the preceding paragraph?

To answer this question we note that for identical values of R the cross-sectional area of a circle is approximately a factor of 7 larger than the area of a triangle with side R (Fig. 8). It follows from (7.8) that the flow rate Q is proportional to the square of the pipe cross section. The square of the area of an equilateral triangle is smaller than the square of the area of a circle by about a factor of 50. Multiplying $3^{1/2}/320$ by 50, we obtain approximately 0.3, which is comparable to the factor $\pi/8 \approx 0.4$ in the case of a circular pipe.

Therefore (7.8) can be improved by writing it in the form

$$Q \sim \rho\,\delta P\,S^2/(\eta l), \tag{7.9}$$

where S is the cross-sectional area of the pipe. Then the proportionality constant in (7.9) is always of order unity regardless of the shape of the pipe cross section.

The above discussion is implicitly taken into account by introducing the *hydraulic diameter* $D = 4S/\Pi$ where Π is the perimeter of the pipe cross section. For a circular cross section we have $D = 2R$. In the case of a cross section of complicated

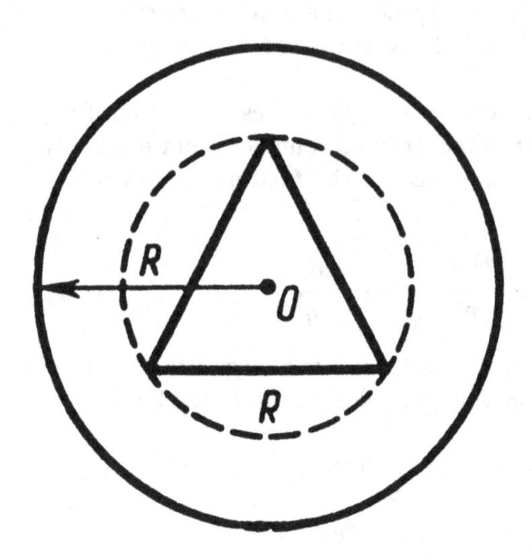

FIG. 8. Circle with radius R equal to the side of an equilateral triangle.

shape the hydraulic diameter determines the flow rate of the fluid. In addition, the hydraulic diameter D determines the Reynolds number $\mathrm{Re} = vD/\nu$ in pipes with complicated shapes.

We emphasize that a large numerical factor can occur in a problem when the required physical quantity (in the problem considered here, the mass flow rate Q) is determined by a parameter raised to a high power (in this problem the parameter is the radius R of the pipe, which is raised to the fourth power in the formula for Q). Then a change in the parameter, say, by a factor of 2, results in a change of the required physical quantity by an order of magnitude or more. From the physical point of view it is completely natural to use the pipe cross section in place of the square of its radius in estimates of the type of (7.8).

Suppose instead of a pipe we have two parallel plates with a small distance h between them. In analogy with (7.6), we can estimate the velocity of the fluid parallel to the surfaces of these plates under a pressure difference δP:

$$v \sim h^2\, \delta P/(\eta l). \tag{7.10}$$

Here, l is the length of a plate and δP is the difference in pressures on the two ends of the plate (δP is assumed to be small so that the Reynolds number $\mathrm{Re} = \rho v h/\eta \ll 1$). We note that for the maximum flow velocity (in the middle of the gap between the plates) the numerical factor in (7.10) is 1/8. Therefore we obtain for the mass flow rate Q of the fluid

$$Q \sim \rho v h b \sim \rho b h^3\, \delta P/(\eta l), \tag{7.11}$$

where $b \gg h$ is the width of the plates.

Comparing (7.11) and (7.9), we conclude that when the linear dimensions of the pipe cross section are different (in this problem $b \gg h$) it is more correct to put in place of (7.9)

$$Q \sim \rho\, \delta P\, S h^2/(\eta l). \tag{7.12}$$

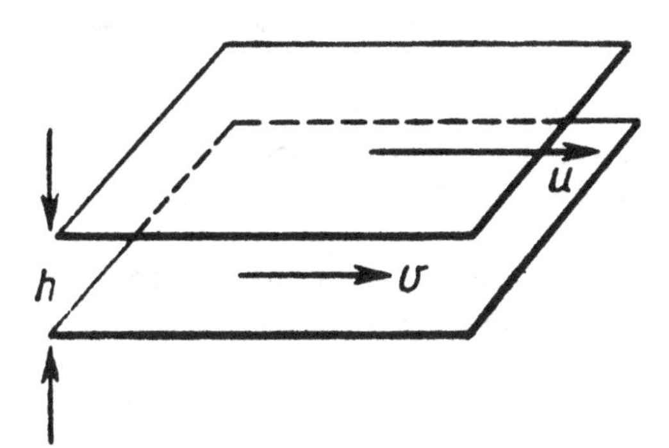

FIG. 9. Flow of a fluid between two parallel plates, one of which moves relative to the other with velocity u.

where h is the smallest linear dimension of the cross section and S is the cross-sectional area of the pipe.

The formula (7.6) is also valid for a porous medium, where now R is the radius of a pore. The quantity $k \sim R^2$ is called the *permeability constant* and it has units of m^2. We then obtain from (7.6)

$$v = k\,\delta P/(\eta l). \tag{7.13}$$

This expression defines the permeability constant exactly, and not as an order of magnitude estimate. Hence we obtain for the mass flow rate $Q = \rho v S$ through a cross section S of a porous medium

$$Q = k\rho\,\delta P\,S/(\eta l) \tag{7.14}$$

(*Darcy's law*). We see that the flow rate is proportional to the first power of the cross-sectional area S, and not S^2, as in (7.9). This follows from the assumption that when a liquid (or gas) flows through a porous medium, the flux through a given pore is independent of the other pores.

Next, suppose we have two parallel plates separated by a small distance h and suppose there is no external pressure difference. Rather, one of the plates moves relative to the other and parallel to it with velocity u (Fig. 9). Then, as in (7.7), the series expansion of the velocity v of the fluid between the plates in the Reynolds number $\mathrm{Re} = \rho u h/\eta \ll 1$ has the form

$$v \sim u[1 + C\rho u h/\eta]. \tag{7.15}$$

Here, C is a numerical factor.

All of the results obtained in this section are valid for both liquids and gases, as long as we can neglect the change in gas density as a result of the pressure gradient. This will be the case if the pipe in which the gas flows is sufficiently short. For large lengths l, when the change of density is significant, it is necessary to integrate the above results with respect to the length of the pipe, assuming that over a small length dl the gas is incompressible and the results obtained above are valid.

We illustrate these ideas for the example of isothermal flow of a viscous gas through a pipe. Let T be the temperature of the gas, l the length of the pipe, and R

its linear dimension in the transverse direction. The pressures on the ends of the pipe are P_1 and P_2. We estimate the mass flow rate of gas Q through the pipe, assuming the gas is ideal.

According to (1.38), the dynamical viscosity is $\eta \sim \sqrt{MT}/\sigma$, where M is the mass of a gas molecule and σ is the cross section for molecule-molecule collisions. Therefore at a given temperature T the dynamical viscosity η does not depend on the gas pressure P or its density ρ.

Over a sufficiently small distance dl the gas can be considered incompressible and we can use Poiseuille's law (7.8). We write this formula in the form

$$Q \sim \frac{\rho R^4}{\eta} \frac{dP}{dl}. \tag{7.16}$$

Here, dP is the pressure difference over a small segment of length dl. We substitute $\rho = MP/T$ for the ideal-gas density into (7.16). Then

$$Q \sim \frac{MR^4}{\eta T} \frac{P\,dP}{dl}. \tag{7.17}$$

We integrate this equation over the length of the pipe l, where the change in the density ρ of the gas is not small. Using the fact that the mass flow rate of the gas Q is constant along the pipe, we have

$$Q \sim \frac{MR^4(P_2^2 - P_1^2)}{\eta T l}. \tag{7.18}$$

If the pressure drop is small so that $P_2 - P_1 \ll P_1, P_2$, then (7.18) reduces to Poiseuille's law (7.16), as expected. If, on the other hand, the pressure drop is very large, so that $P_2 \gg P_1$, then it is evident from (7.18) that the mass flow rate of the gas Q is proportional to the square of the pressure P_2. We must have $\mathrm{Re} \ll 1$, however.

Similarly, for the case of gas flow through a porous medium we replace R^4 in (7.18) by kS, where S is the cross-sectional area of the porous medium and k is the permeability constant [as was done in obtaining Darcy's law (7.14)]. Then we obtain

$$Q \sim \frac{kMS}{\eta T l} (P_2^2 - P_1^2). \tag{7.19}$$

Hence *when the pressure drop is large Darcy's law is quadratic in this pressure drop.*

Problem 1. Show that if the porosity ε, defined as the relative volume of all cavities in a given volume of the medium, is small, i.e., $\varepsilon \ll 1$, then the permeability constant k is proportional to ε^3.

Problem 2. Estimate the mass flow rate of a viscous gas in the case of adiabatic flow through a long pipe under a large pressure drop ΔP. Assume that the Reynolds number is small. Show that the mass flow rate Q is of order

$$Q \sim (\Delta P)^{(3+\gamma)/(2\gamma)},$$

where $\gamma = c_P/c_V$ is the adiabatic exponent.

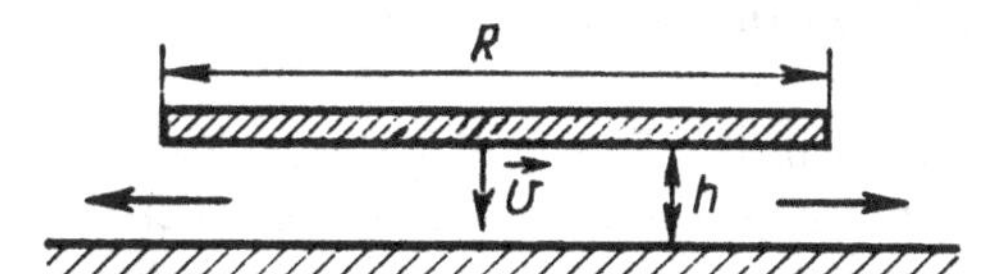

FIG. 10. Extrusion of a liquid from a gap of thickness h by a horizontal plate with linear dimension R under the force of gravity on the plate.

Problem 3. A wide pipe filled with liquid begins to move along its axis with a small velocity u, dragging along the liquid with it. Show that the frictional force per unit area acting on the inner surface of the pipe decreases with time t according to the equation $F_x \sim u(\rho\eta/t)^{1/2}$.

Problem 4. A horizontal plate of mass m and linear transverse dimension R is placed in a liquid at a small distance $h \ll R$ from the bottom (Fig. 10). The plate sinks under gravity, extruding the liquid from the gap. Show that the velocity v at which the plate sinks is of order

$$v \sim mgh^3/(\eta R^4).$$

7.2. Motion of a body in a fluid

We estimate the drag force on a solid body with characteristic linear dimension R moving with a constant velocity u in a viscous fluid with dynamical viscosity η. The Reynolds number is assumed to be small in comparison with unity.

For small velocity u (corresponding to small Reynolds number) the drag force F can be expanded in a Taylor series in the dimensionless parameter Re, which is proportional to the velocity of the body u. In the linear approximation the force F is proportional to the first power of u. The proportionality constant is constructed from R and η:

$$F \sim \eta u R. \tag{7.20}$$

For a sphere of radius R the proportionality constant in this equation is equal to 6π, and then (7.20) is called **Stokes's law**.

Including the next term of the Taylor series in $\mathrm{Re} = \rho u R/\eta$, we obtain

$$F \sim \eta u R(1 + C \cdot \mathrm{Re}). \tag{7.21}$$

A numerical calculation shows that $C = \frac{3}{8}$ for a sphere of radius R.

Instead of a sphere, suppose we have a cylinder of radius R moving perpendicular to its axis with velocity u. Then the problem is to estimate the drag force acting on a unit length of the cylinder. Dividing the right-hand side of (7.20) by the characteristic length of the problem, i.e., R, we obtain

$$F \sim \eta u. \tag{7.22}$$

We see that the radius of the cylinder drops out of the expression for the drag force; it appears only in the next order correction in the Reynolds number.

Stokes's law (7.20) also holds for the motion of a gas bubble in a liquid. The only difference is that the numerical factor is 12π instead of 6π. Unlike the case of a solid body, when a liquid flows around a gas bubble the tangential velocity of the liquid does not have to vanish on the surface of the bubble. Only the normal component of the velocity is zero. The change in shape of the bubble due to motion in the liquid is neglected, which is valid when the pressure inside the bubble is sufficiently high.

It follows from the above results that the viscosity of the fluid is the determining factor in the drag force. We consider the distance from a moving body for which the viscosity of the fluid significantly affects the streamlines and the limiting distance for which the fluid can be treated as ideal.

Suppose a body with linear dimension R moves in a fluid with constant velocity $\mathbf{u}$. We consider large distances $r \gg R$ from the body. In a coordinate system moving with the body, the velocity of the fluid at a given point in space is $\mathbf{u} + \mathbf{v}$, where $u \gg v$ when $r \gg R$. Then the left-hand side of (7.4) is roughly $(\mathbf{u}\cdot\nabla)\mathbf{v} \sim uv/r$, whereas the viscous term on the right-hand side of (7.4) is of order $\nu v/r^2$. Comparing these two quantities with one another, we conclude that at distances $r \gg R$, $r \ll \nu/u$ the nonlinear term $(\mathbf{u}\cdot\nabla)\mathbf{v}$ in the Navier–Stokes equation can be neglected and we have linear viscous flow described by (7.5). Hence the dependence $v \sim r^{-3}$ given by (6.6) is not correct because the boundary conditions at the surface of the body in the viscous case state that both the normal and tangential components of the velocity must vanish, whereas in the case of an ideal fluid only the normal component of the velocity must vanish at the surface.

How does v fall off with r in the case of viscous flow? Because the curl of any gradient is equal to zero, we obtain from (7.5) $\text{curl}(\Delta\mathbf{v}) = 0$. The function $v \sim 1/r$ is a solution of Laplace's equation $\Delta\mathbf{v} = 0$ and is therefore also a solution of the equation $\text{curl}\,(\Delta\mathbf{v}) = 0$. In contrast to the case of an ideal fluid, this solution does not lead to an infinite quantity of fluid carried along by the body, as discussed in Sec. 6.1, since the viscosity of the fluid changes this quantity. Therefore in the case of a viscous fluid the velocity of the fluid v falls off with distance from the moving body as $1/r$ for distances $R \ll r \ll \nu/u$. According to (7.5), the excess pressure caused by the motion of the body falls off as r^{-2} for these distances.

At larger distances $r \sim \nu/u$ the term $(\mathbf{v}\cdot\nabla)\mathbf{v}$ in the Navier–Stokes equation (7.4) cannot be neglected, in spite of the fact that the Reynolds number $\text{Re} = uR/\nu \ll 1$. However, the Navier–Stokes equation does not really become nonlinear in this case, since $v \ll u$ and the term $(\mathbf{v}\cdot\nabla)\mathbf{v}$ can be replaced by $(\mathbf{u}\cdot\nabla)\mathbf{v}$. Then (7.4) takes the form

$$(\mathbf{u}\nabla)\mathbf{v} = -\frac{1}{\rho}\nabla P + \nu\Delta\mathbf{v} \quad (r \gg R). \tag{7.23}$$

This equation is called the **Oseen equation.**

For still larger distances $r \gg \nu/u$ we can neglect the viscous term in (7.23) and the fluid can be assumed to be ideal. However, we will see in the next section that this conclusion is correct almost everywhere except in a narrow region far behind the moving body.

Up to now we have considered steady viscous flow around the body. We now consider the process of relaxation of the viscous motion. Suppose a body with linear dimension R moves in a fluid and the Reynolds number is small. We estimate the time required for the fluid to return to a state of rest after the body stops moving.

When the Reynolds number is small, the total derivative $d\mathbf{v}/dt$ on the left-hand side of (7.3) can be replaced by the partial derivative $\partial\mathbf{v}/\partial t$ [see the discussion after

(7.4)]. Relaxation is determined by the second term on the right-hand side of (7.3). Hence we find from (7.3) that the damping of the velocity in time is determined qualitatively by the equation

$$\frac{\partial \mathbf{v}}{\partial t} \sim \nu \Delta \mathbf{v}. \tag{7.24}$$

The size of the region of fluid perturbed by the motion of the body is of the order of the linear dimension of the body R. Therefore $\Delta \mathbf{v} \sim \mathbf{v}/R^2$. Furthermore, on the left-hand side of (7.24) we can set $\partial \mathbf{v}/\partial t \sim \mathbf{v}/\tau$, where τ is the relaxation time of the motion of the viscous fluid (the time required for the fluid to reach the state of rest). Substituting these estimates into (7.24), we find

$$\tau \sim R^2/\nu. \tag{7.25}$$

We see that *the time τ required for the fluid to reach the state of rest after the body stops moving is independent of the velocity of the body.* This fact is a consequence of the linearity of the Navier–Stokes equation for small Reynolds numbers $\mathrm{Re} = vR/\nu \ll 1$.

Problem 1. A particle of linear dimension R is placed in a fluid moving with velocity v. The density of the particle is comparable to the density of the fluid, and therefore gravitational effects on the particle can be neglected. Show that the particle acquires the velocity of the fluid after being carried along by the fluid over a distance of order uR^2/ν, where ν is the kinematic viscosity of the fluid.

Problem 2. A liquid layer of thickness h rolls down an inclined trough under gravity. The linear transverse dimension of the trough is b. Show that the flow rate of the liquid Q is of order

$$Q \sim \rho g h^3 b/\nu,$$

where ν is the kinematic viscosity of the liquid, g is the acceleration of gravity, and ρ is the density of the liquid.

7.3. Laminar wake

A *laminar wake* is a narrow region far behind a moving body where the viscosity of the fluid cannot be neglected, even though the distance from the body is large $x \gg \nu/u$ [see the discussion of the preceding section after (7.23)]. Here, u is the stream velocity of the fluid, which flows around a body at rest.

We estimate first the width y of the laminar wake as a function of the distance x from the body. We put the origin of our coordinate system at the center of the body (Fig. 11). Let R be the linear dimension of the body. We consider distances $x \gg R$ far behind the body. We will see that the width of the wake satisfies $y \ll x$, but $y \gg R$.

We write the velocity of the fluid at a point in the wake in the form $\mathbf{u} + \mathbf{v}$, where $v \ll u$. We have $v_x < 0$ in the wake, since the fluid in the wake flows slightly more slowly than outside it.

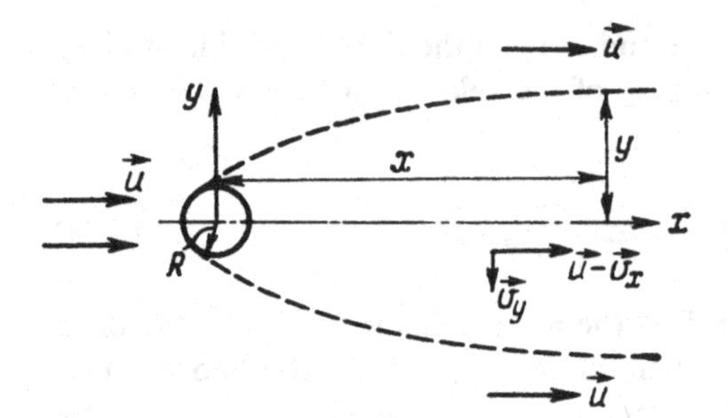

FIG. 11. Laminar wake behind a moving body.

Using the incompressibility equation (6.1), we compare the velocity components v_x and v_y:

$$v_y \sim \frac{y}{x} v_x \ll v_x. \tag{7.26}$$

Taking the y component of the Oseen equation (7.23) and estimating the various terms in this equation using the fact that $y \ll x$, we find

$$uv_y/x \sim P/(\rho y) \sim \nu v_y/y^2. \tag{7.27}$$

Setting the first term in (7.27) equal to the third term, we obtain a relation between y and x:

$$y \sim \left(\frac{\nu x}{u}\right)^{1/2} \sim \left(\frac{x}{R}\right)^{1/2} \frac{R}{\mathrm{Re}^{1/2}}. \tag{7.28}$$

According to (7.28), the condition $y \ll x$ implies that $ux/\nu \gg 1$. As we have seen above in the discussion of (7.23), outside the laminar wake the viscosity of the fluid can be neglected at these distances and the fluid can be treated as ideal.

We note that the condition $ux/\nu \gg 1$ is consistent with the requirement of small Reynolds number $\mathrm{Re} = uR/\nu \ll 1$ if x is sufficiently large in comparison with R. Hence the region $x \gg \nu/u \gg R$ is the region of the laminar wake. Its width is given by (7.28), where $y \gg R$.

The quantity P in (7.27) is the difference in pressure ΔP between the wake and the unperturbed fluid. Comparing the second and third terms in (7.27) and using (7.26), we express P in terms of v_x:

$$P \sim \rho \nu v_y/y \sim \rho \nu v_x/x. \tag{7.29}$$

Taking the x component of (7.23),

$$uv_x/x \sim \nu v_x/y^2. \tag{7.30}$$

Hence we again obtain (7.28). The term in (7.23) involving the pressure is of order $P/\rho x$, which, according to (7.29), is roughly $\nu v_x/x^2$. Since $y \ll x$, this term is small in comparison with the terms in (7.30), and it can be neglected in the x component (7.30) of the vector Oseen equation (7.23).

We note that in obtaining (7.27) from (7.23) we replaced $(\mathbf{u} \cdot \nabla)$ on the left-hand side of (7.23) by u/x. In the exact Navier–Stokes equation (7.4) in a coordinate system moving with the body, the left-hand side contains the operator $(\mathbf{u} + \mathbf{v})\nabla$. The second term in this expression is roughly v_y/y. Therefore we actually assumed that

$$v_y \ll \frac{y}{x} u, \tag{7.31}$$

when we replaced $(\mathbf{u}\cdot\nabla)$ by u/x. The validity of this inequality is proven below, where the velocity components v_x and v_y in the laminar wake are estimated.

We now estimate the typical deviations v_x and v_y of the velocity in a laminar wake from the stream velocity v. The velocities v_x and v_y cannot be estimated directly from the Navier–Stokes equation, since this equation is linear in v. To solve the problem we use the result (7.20) for the drag force on a body moving in a fluid: $F \sim \eta u R$. Here, R is the linear dimension of the body. According to Newton's third law, this force must be equal to the reaction force of the body on the fluid. We consider a sphere a large radius x such that it crosses the laminar wake far behind the body. The Stokes force F is equal to the difference between the forces acting in the laminar wake and in the symmetrically located region in front of the body. The forces acting in front of and behind the body cancel out in the other directions. Therefore we have

$$F = \int\int [P_0 + P + \rho(\mathbf{u}+\mathbf{v})^2] dy\, dz - \int\int [P_0 + \rho u^2] dy\, dz. \tag{7.32}$$

The integrand in (7.32) involves the sum of the static pressure ($P_0 + P$ in the laminar wake and P_0 in front of the body) and the dynamic pressure [$\rho(\mathbf{u} + \mathbf{v})^2$ in the laminar wake and ρu^2 in front of the body]. As before, the x axis is chosen in the direction of the velocity vector $\mathbf{u}$. Neglecting quadratic terms in v in (7.32), we can write

$$(\mathbf{u}+\mathbf{v})^2 - \mathbf{u}^2 \approx 2\mathbf{u}\mathbf{v} = 2uv_x.$$

The terms involving P_0 in (7.32) cancel one another out. Hence we obtain

$$F = \int\int (P + 2\rho u v_x) dy\, dz. \tag{7.33}$$

The excess static pressure P in the laminar wake is given by (7.29). Since $ux \gg \nu$ in the laminar wake, we see that the term in (7.33) involving the static pressure P is negligibly small in comparison with the term $\rho u v_x$. Therefore, keeping only the second term on the right-hand side of (7.33), we find

$$F \sim \rho u v_x y^2. \tag{7.34}$$

We assume that the motion of the fluid is axially symmetric and therefore the effective area of integration over the laminar wake is replaced by y^2.

Setting (7.34) to the Stokes force $\eta u R$, we obtain an estimate for the velocity v_x:

$$v_x \sim \nu R / y^2. \tag{7.35}$$

Substituting the estimate (7.28) for the width of the wake y into (7.35), we finally obtain for the velocity

$$v_x \sim \frac{R}{x} u. \tag{7.36}$$

As noted above, this expression is correct in the laminar wake where $x \gg \nu/u$. However, as we saw above in the derivation of the Oseen equation (7.23), this estimate is also correct in the region $R \ll x \lesssim \nu/u$.

Because $x \gg R$, it follows from (7.36) that we indeed have $v_x \ll u$, which was used above.

From (7.26) and (7.28), we obtain for the transverse component of the velocity v_y

$$v_y \sim \left(\frac{\nu}{ux^3}\right)^{1/2} Ru. \tag{7.37}$$

We see that as x increases the transverse velocity component v_y falls off more rapidly than the longitudinal velocity component v_x. This fact determines the existence of the laminar wake.

We check whether the inequality (7.31), which was used above, is actually satisfied. From (7.36) we have

$$\frac{v_y}{yu/x} \sim \frac{v_x}{u} \sim \frac{R}{x} \ll 1, \tag{7.38}$$

which was to be shown.

Substituting (7.36) into (7.29), we obtain an estimate for the excess pressure P in the laminar wake (P is the excess above the unperturbed pressure P_0):

$$P \sim \eta u R / x^2. \tag{7.39}$$

As in the derivation of (7.36), this estimate also holds in the region $R \ll x \lesssim \nu/u$.

Physically, the existence of a laminar wake, where the viscosity cannot be neglected at large distances $x \gg \nu/u$, is a consequence of the fact that the derivative of the velocity $\mathbf{v}$ with respect to the transverse coordinate y is large inside the laminar wake.

Problem 1. Show that the flow of fluid inside a laminar wake is rotational, i.e., curl $\mathbf{v} \neq 0$.

7.4. Energy absorption in a viscous fluid

We estimate the power absorbed in a viscous fluid when a solid body vibrates in the fluid. Let the linear dimension of the body be R and the vibration frequency be ω. Also, let u be the amplitude of the oscillating velocity of the body. As in the preceding sections of this chapter, we assume that the Reynolds number is small.

We turn to the general Navier–Stokes equation (7.3). Because the Reynolds number is small, the term $(\mathbf{v}\cdot\nabla)\mathbf{v}$ can be neglected. The term ∇P on the right-hand side of (7.3) cannot be large in comparison with $\rho\, \partial\mathbf{v}/\partial t$, because otherwise the problem would be steady. As we saw in the case of viscous relaxation, the damping is determined by the second term on the right-hand side of (7.3). Therefore we obtain the equation

$$\frac{\partial \mathbf{v}}{\partial t} \sim \nu \Delta \mathbf{v}, \tag{7.40}$$

describing qualitatively the damping of the vibrational motion [see also (7.24)]. Here, ν is the kinematic viscosity of the fluid.

Estimating the left-hand side of (7.40) as ωv and the right-hand side as $\nu v/\delta^2$, where δ is the typical distance over which the motion of the viscous fluid is oscillatory in nature, we obtain $\omega v \sim \nu v/\delta^2$, and hence

$$\delta \sim (\nu/\omega)^{1/2}. \tag{7.41}$$

At large distances $r \gg \delta$ the motion, although unsteady, is not oscillatory with the frequency ω of forced vibrations created by the body.

We estimate the contribution to the dissipation due to the transformation of the kinetic energy of the vibrating body into heat by means of viscosity. This process takes place in a layer of thickness δ surrounding the body. The kinetic energy of the fluid per unit volume is of order ρv^2, where ρ is the density of the fluid and v is its velocity. The rate of change of this energy is of order

$$j \sim \rho v \frac{\partial v}{\partial t}. \tag{7.42}$$

This is therefore the power absorbed per unit volume of fluid.

For vibrational motion with frequency ω we have

$$\partial v/\partial t \sim \omega v,$$

and therefore we obtain for j

$$j \sim \rho \omega v^2. \tag{7.43}$$

The total absorbed power is obtained from (7.43) by multiplying it by the thickness δ of the layer over which the vibrational motion of the fluid extends and by the surface area of the body (of order R^2). Let the total dissipated power be $\dot{E} = \partial E/\partial t$. We find

$$\dot{E} \sim j\,\delta R^2 \sim \rho v^2 R^2 (\nu\omega)^{1/2}. \tag{7.44}$$

Assume first that $\delta \ll R$. The dissipation in a thin layer surrounding the vibrating body is given by (7.44). We now must estimate the contribution to the dissipation from a region with linear dimension of order R. In this region the motion of the fluid is not oscillatory, and the fluid velocity falls off with distance as $\exp(-r/\delta)$. We assume that the vibration amplitude v/ω is small in comparison with the size of the body, i.e., $v \ll \omega R$. Because of the exponentially small factor $\exp(-R/\delta)$, the contribution from the region of linear dimension R is obviously very small.

If $\delta \sim R$ the total dissipated power is obtained from (7.44)

$$\dot{E} \sim \rho v^2 R \nu. \tag{7.45}$$

If $\delta \gg R$ the energy of the fluid is still absorbed mainly in a region with linear dimension of order R, since the velocity of the fluid decreases rapidly with larger distances. According to (7.40), the derivative $\partial v/\partial t$ can be estimated as $\nu v/R^2$. Then using (7.42) we obtain the dissipated power $\dot{E}$

$$\dot{E} \sim jR^3 \sim \rho v \frac{\partial v}{\partial t} R^3 \sim \rho v^2 R \nu, \tag{7.46}$$

which, as expected, is consistent with the estimate (7.45).

Therefore we conclude that when $\delta \ll R$ the energy is absorbed in a thin surface layer of thickness δ surrounding the vibrating body, whereas when $\delta \gtrsim R$ energy is absorbed in a region whose linear dimension is of the order of the linear dimension of the body R.

We now consider energy absorption in the case of propagation of gravity waves on the surface of a liquid (see Sec. 6.2). The power dissipated per unit volume is given by (7.42). Using (7.40), we obtain

$$j \sim \rho v^2 \nu / \lambda^2. \tag{7.47}$$

The wavelength λ serves as the characteristic unit of length analogous to the linear dimension R of the body in the preceding problem. Indeed, as we saw in Sec. 6.2, a layer of liquid with thickness of order λ participates in the vibrational motion in a propagating gravity wave.

The *damping constant* γ is defined as the ratio of the dissipated power per unit volume to the energy density ρv^2 of the fluid. Using (7.47), we obtain the estimate

$$\gamma \sim \nu / \lambda^2. \tag{7.48}$$

Substituting (6.12) into (7.48), we obtain an estimate for γ involving the frequency of the wave ω

$$\gamma \sim \nu \omega^4 / g^2. \tag{7.49}$$

Therefore *the damping constant of a gravity wave is proportional to the fourth power of its frequency.*

We discuss the condition under which (7.49) is applicable. The fraction of the energy of the wave transformed into heat after a period of vibration must be small (otherwise we do not have wave motion), which implies that

$$jT \ll \rho v^2, \tag{7.50}$$

where T is the period of vibration, which is of order ω^{-1}. Substituting (7.47) into (7.50), this inequality can be rewritten in the form

$$\rho v^2 \nu / (\omega \lambda^2) \ll \rho v^2, \quad \text{or} \quad \nu \ll \omega \lambda^2. \tag{7.51}$$

Expressing λ in terms of ω with the help of (6.12), we finally obtain

$$\nu \ll g^2 / \omega^3. \tag{7.52}$$

This condition is a restriction on the viscosity of the liquid. If the viscosity is too large then gravity waves cannot exist on the surface of the liquid because the energy of the wave is dissipated into heat so rapidly that the motion is aperiodic.

These results hold for the case of a deep liquid, when the depth satisfies $h \gg \lambda$. We consider now the opposite limiting case $h \ll \lambda$. Then in the general relation (7.48), which holds for any dispersion law, it is necessary to substitute the value of λ from (6.16) for the case of a shallow liquid:

$$\gamma \sim \nu\omega^2/(gh). \tag{7.53}$$

We see that in this case the damping constant is proportional to the square of the wave frequency ω, and not the fourth power, as in the case of a deep liquid.

The condition (7.51) that the wave must not damp out after one or two periods of vibration then takes the form

$$\nu \ll gh/\omega. \tag{7.54}$$

When $h \sim \lambda$ the conditions (7.52) and (7.54) coincide, as they must:

$$\nu \ll (gh^3)^{1/2}. \tag{7.55}$$

Then from (7.48) the damping constant in this case has the form

$$\gamma \sim \nu/h^2. \tag{7.56}$$

We note that the general condition that a wave must not damp out in one or two periods of vibration (i.e., that the vibrations must not be aperiodic) is equivalent to the condition that the imaginary part of the frequency must be small in comparison with the real part, i.e., $\gamma \ll \omega$. This question was considered in detail in Chap. 3 [see (3.31)].

The result (7.53) is correct when $\delta \gg \lambda$, where

$$\delta \sim (\nu/\omega)^{1/2}$$

is the thickness of the layer over which energy is absorbed.

If $\delta \lesssim \lambda$ then the power $\dot{j}$ dissipated per unit volume is given by (7.43). Suppose first that $\delta \ll h \ll \lambda$. Then the dissipated power E per unit area of the bottom is of order $j\delta$. Dividing it by the kinetic energy $\rho v^2 h$ of the vibrating shallow liquid per unit area, we obtain the damping constant γ for the case $\delta \ll h$, i.e., $\nu \ll \omega h^2$:

$$\gamma \sim (\nu\omega)^{1/2}/h. \tag{7.57}$$

The damping is determined by friction of the liquid against the bottom. The result (7.57) is correct when $\gamma \ll \omega$, i.e.,

$$\nu \ll \omega h^2, \tag{7.58}$$

which was already assumed above.

If $h \lesssim \delta \lesssim \lambda$ then $\gamma \sim \omega$ and waves cannot propagate in a shallow liquid: the motion damps out aperiodically.

Problem 1. A long cylinder of radius R vibrates in the direction perpendicular to its axis with frequency ω and velocity amplitude u and

$$(\nu/\omega)^{1/2} \ll R.$$

Show that the drag force of the liquid acting on a unit length of the cylinder is of order

$$F \sim Ru(\omega\eta\rho)^{1/2},$$

where ρ is the density of the fluid and η is its dynamical viscosity.

Chapter 8

Turbulence

Turbulent flow occurs in a liquid or gas at large Reynolds numbers

$$\mathrm{Re} = \frac{uR}{\nu} \gg 1,$$

where u is the fluid velocity, R is the linear dimension of a body placed in the moving fluid, and ν is the kinematic viscosity of the fluid. If the dimensions of the body are different in different directions, then R is taken to be the transverse dimension with respect to the direction of the flow.

Turbulent motion is characterized by vortex motion with a very wide range of length scales. The largest vortices have radii of order R, since this is the largest linear dimension in the problem. The stream velocity u is also the typical velocity of fluid particles in the largest vortices [neglecting logarithmic factors which become significant in a more detailed treatment; see (9.28)].

We rewrite the Navier–Stokes equation (7.3) in the form

$$\frac{\partial \mathbf{v}}{\partial t} + (\mathbf{v}\nabla)\mathbf{v} = -\frac{1}{\rho}\nabla p + \nu \Delta \mathbf{v}. \tag{8.1}$$

Since

$$(\mathbf{v}\nabla)\mathbf{v} \sim u^2/R,$$

where u is the stream velocity, while

$$\nu \Delta \mathbf{v} \sim \nu u/R^2,$$

the ratio of the first expression to the second is of order $\mathrm{Re} = uR/\nu \gg 1$. Therefore the viscous term $\nu\Delta\mathbf{v}$ in the Navier–Stokes equation (8.1) can be neglected for the largest vortices and so *the flow is inviscid (ideal) in a vortex of the maximum size R*. Over distances of this order the kinetic energy of the fluid is not transformed into heat.

The only transfer of energy that can take place is a transfer of kinetic energy from the larger vortices into the smaller ones. As the size of a vortex decreases the Reynolds number associated with it also decreases, since it is proportional to the vortex size. In addition, we will see below that as the size of a vortex decreases the corresponding velocity of vortex motion also decreases, which further decreases the Reynolds number.

When the Reynolds number reaches a value of order unity (more exactly, the critical Reynolds number $\mathrm{Re}_{\mathrm{cr}}$; see the more detailed discussion in Sec. 8.1), the term $\nu\Delta\mathbf{v}$ in (8.1) can no longer be neglected and then the kinetic energy of the vortex motion can be transformed into heat, i.e., we have energy dissipation.

Let ε be the power absorbed per unit mass of the fluid. This power is transformed from vortices with linear dimension R to smaller and smaller vortices until we reach vortices with linear dimensions such that kinetic energy is transformed into heat. The quantity ε can be determined from dimensional considerations using the velocity u, linear dimension R, and density of the fluid ρ. The kinematic viscosity cannot appear in the relation for ε because the motion in a vortex of linear dimension R is ideal. Since the dimensions of ε are $\mathrm{J/(kg\ s)} = \mathrm{m^2/s^3}$, we obtain

$$\varepsilon \sim u^3/R \tag{8.2}$$

(the density ρ does not appear because of dimensional considerations).

We see that the rate of energy dissipation in turbulent flow is much larger than in laminar flow. In the latter case it is roughly $\nu u^2/R^2$ [see (7.46)], which is smaller than (8.2) by a factor of Re.

The typical relaxation time of energy dissipation τ can be obtained from the kinetic energy per unit mass of fluid $E \sim u^2$ as $\tau \sim E/\varepsilon$. Therefore we find from (8.2)

$$\tau \sim R/u. \tag{8.3}$$

We see that this time is of the order of a period of oscillation of the largest vortices in turbulent flow. Therefore energy dissipation itself is a relatively fast process, but the time required for dissipation to occur is much larger, since it is determined by numerous processes of energy exchange between vortices of different sizes. The sizes of the vortices both increase and decrease as a result of the exchange process and kinetic energy is transformed into heat only in the smallest vortices.

Similarly, we can use dimensional analysis to estimate the pressure acting on a body moving with respect to the fluid in the case of large Reynolds number:

$$P \sim \rho u^2. \tag{8.4}$$

The viscosity ν does not appear in (8.4) because the fluid is ideal. According to (8.4), the drag force is of order $F \sim \rho u^2 R^2$, which is larger than the Stokes force (7.20), corresponding to laminar flow, by a factor of Re.

8.1. Fully developed turbulence

The critical Reynolds number $\mathrm{Re}_{\mathrm{cr}}$ at which the flow changes from laminar to turbulent depends on the detailed configuration of the system; it is always larger than unity. The picture discussed above of vortices of different sizes occurs when $\mathrm{Re} \gg \mathrm{Re}_{\mathrm{cr}}$ and then we speak of *fully developed turbulence.*

We estimate the velocity of fluid particles in vortices of different sizes in the case of fully developed turbulence. We consider vortex motion with a characteristic length l in the interval $l_0 \ll l < R$, where l_0 is the size of the vortex for which the Reynolds number is of order unity. As noted above, the quantity R is the maximum size of a vortex. Let v_l be the velocity of a vortex of size l.

The quantity v_l depends only on the characteristics of the vortex, including the dissipated power ε transferred from the larger vortices to the smaller ones. In addition, v_l can depend on the size l of the vortex and on the density of the fluid ρ. It cannot depend on the viscosity ν because vortex motion is inviscid. As in the derivation of (8.2), we obtain from dimensional considerations

$$v_l \sim (\varepsilon l)^{1/3} \sim \left(\frac{l}{R}\right)^{1/3} u. \tag{8.5}$$

Here, we have used the estimate (8.2) for ε. We see that the velocity v_l decreases with decreasing l. The dependence (8.5) is called the **Kolmogorov–Obukhov law.**

We introduce the Reynolds number

$$\mathrm{Re}_l = v_l l/\nu$$

for a vortex with linear dimension l. We estimate the smallest value l_0, for which $\mathrm{Re}_{l_0} \sim 1$ and kinetic energy of the vortex motion is dissipated into heat. Using (8.5), we find

$$l_0 \sim \left(\frac{\nu}{u}\right)^{3/4} R^{1/4} \sim \frac{R}{(\mathrm{Re})^{3/4}} \ll R. \tag{8.6}$$

Here,

$$\mathrm{Re} = uR/\nu$$

is the Reynolds number for the largest vortex.

Substituting (8.6) into (8.5) for $l \sim l_0$, we obtain the velocity of fluid particles in the smallest vortex:

$$v_{l_0} \sim \frac{u}{(\mathrm{Re})^{1/4}} \ll u. \tag{8.7}$$

We consider now the flow of a fluid with fully developed turbulence over distances $l \ll l_0$, where vortex motion does not occur. In this case the fluid velocity is small and the nonlinear term in the velocity in the Navier–Stokes equation (8.1) can be neglected. Then the solution of the linear hydrodynamic equation can be expanded in a Taylor series in the small parameter $l/l_0 \ll 1$. Hence we conclude that when $l \ll l_0$ the velocity v_l is proportional to the first power of the scale l of the laminar flow. The proportionality constant in this dependence is found from the condition that $v_l \sim v_{l_0}$ when $l \sim l_0$. Therefore we find for $l \ll l_0$

$$v_l \sim \frac{l}{l_0} v_{l_0} \sim \frac{l}{R} (\mathrm{Re})^{1/2} u. \tag{8.8}$$

As a simple example of a problem involving fully developed turbulence, we consider the flow of a liquid from a pipe into an infinite space filled with the same liquid (a so-called submerged jet). Let u be the velocity of the liquid in the pipe and R be the radius of the pipe. We will be interested only in distances which are large in comparison with the pipe radius R.

The propagation of a submerged jet is shown schematically in Fig. 12. We let y be the width of the turbulent flow at large distances $x \gg R$ from the pipe opening. Because there is no other scale of length in the problem (the linear dimension R is

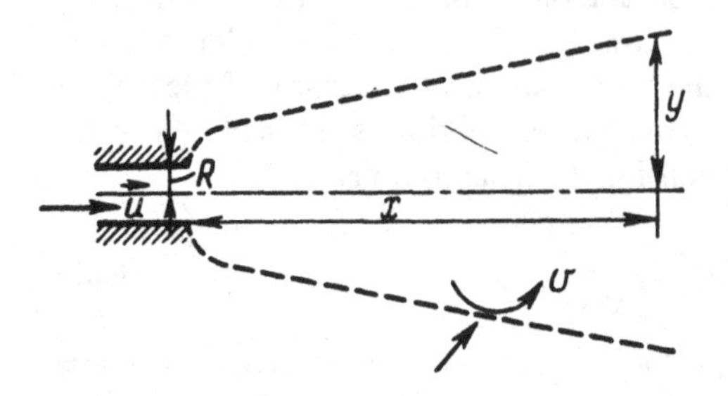

FIG. 12. Propagation of a turbulent submerged jet from a pipe of radius R; here v is the velocity of turbulent motion inside the jet and the dashed lines show the boundary of the jet.

not important at large distances x, while the length constructed with the help of the kinematic viscosity ν and the velocity u cannot appear in the problem, since when $\mathrm{Re} \gg 1$ the flow of the liquid is ideal), y and x must be proportional to one another:

$$y \sim x. \tag{8.9}$$

The proportionality constant in this dependence is universal for a given shape of the pipe opening. For a circular opening, for example, it follows from experimental data that the angular width of the cone (see Fig. 12) into which the jet propagates is equal to about 25°. We note that at the cone vortex the angular width of the cone is much larger.

The momentum current of fluid is the same through each cross section of the cone perpendicular to the x axis. It can be estimated as $\rho v \cdot v \cdot y^2$. Here ρ is the density of the fluid and v is the velocity of the jet at the given cross section at distance x from the pipe opening. Using (8.9), we conclude that the velocity v is inversely proportional to x. The proportionality constant can be estimated using the fact that when $x \sim R$ the velocity should be approximately equal to the velocity u of the liquid from the pipe opening. Hence we finally obtain

$$v \sim \frac{R}{x} u. \tag{8.10}$$

We estimate the flow rate Q of liquid through a cross section of the cone at a distance $x \gg R$ from the pipe opening. By definition we have

$$Q \sim \rho v y^2. \tag{8.11}$$

Substituting (8.9) and (8.10) into (8.11), we find that Q is proportional to x. The proportionality constant is found in the same way as in (8.10):

$$Q \sim \frac{x}{R} Q_0. \tag{8.12}$$

Here, Q_0 is the flow rate inside the pipe. It follows from experiment that for a circular pipe of radius R the proportionality constant in (8.12) is approximately equal to 1.5.

We see from (8.12) that the flow rate Q increases with increasing x. This is explained by the fact that liquid flows into the cone containing the turbulent jet through the surface of the cone from the surrounding liquid, in which the flow is laminar. In Fig. 12 streamlines are shown near the boundary of the jet (shown by the dashed lines).

The Reynolds number

$$\mathrm{Re}=vy/\nu \gg 1$$

is independent of x, according to (8.10). But this does not mean that turbulent flow remains at arbitrarily large distances. We have neglected the back pressure P_0 of the liquid in the surrounding volume in comparison with the kinematic pressure ρv^2 of the turbulent jet [see (8.4)]. Therefore the above discussion of the flow of a submerged jet is valid when

$$R \ll x \ll (\rho u^2/P_0)^{1/2} R. \tag{8.13}$$

We consider a pipe opening in the shape of a narrow slit of width h. In this case the boundaries of the turbulent jet are planes. As before, x is the distance from the slit to a plane perpendicular to the direction of the jet and y is the width of the jet at distance x. As in (8.9), we conclude that y is proportional to x. However, the proportionality constant will obviously be different from the case of a circular pipe. The angular width of the jet turns out to be about 25°, as in the case of a circular pipe.

The momentum current density $\rho v \cdot v$ creates a momentum current $\rho v^2 y$ per unit length of the slit. From the fact that the momentum current is constant, it follows that v is inversely proportional to $x^{1/2}$. Instead of (8.10) we have

$$v \sim \left(\frac{h}{x}\right)^{1/2} u. \tag{8.14}$$

Here, u is the velocity of the liquid leaving the slit.

The flow rate Q per unit length of the slit is roughly $\rho v y$ at a distance x from the slit. Therefore Q increases with x as $x^{1/2}$. Instead of (8.12) we have

$$Q \sim (x/h)^{1/2} Q_0. \tag{8.15}$$

Here, Q_0 is the flow rate per unit length inside the slit.

Problem 1. Show that the average velocity of laminar flow outside a submerged jet is small in comparison with the typical velocity of turbulent flow inside the jet at the same distance from the pipe.

Problem 2. Show that after a small time τ the velocity of a given fluid particle in the case of fully developed turbulence changes by a quantity of order $(\varepsilon\tau)^{1/2}$, whereas the velocity at a given point of space changes by a quantity of order $(\varepsilon u\tau)^{1/3}$.

8.2. Turbulent wakes

A laminar wake far behind a body moving in a fluid at small Reynolds numbers was considered in Sec. 7.3. Here, we consider the case of large Reynolds numbers $\mathrm{Re} = uR/\nu \gg 1$, where R is the linear dimension of the body, u is the stream velocity of the fluid, and ν is the kinematic viscosity of the fluid. We consider the fluid motion far behind the body. A *turbulent wake* is a region of turbulent motion of the fluid.

We estimate the width y of a turbulent wake. The boundary between the turbulent and laminar regions is the boundary of the turbulent wake. As we saw in the preceding section in the discussion of a submerged jet, fluid can only flow into the turbulent region, but cannot flow out of it. This means that the streamlines are discontinuous on the boundary of a turbulent wake, like the streamlines shown in Fig. 12. In addition, on the turbulent region side of the boundary the streamlines are parallel to the boundary. We use this picture to determine the shape of the boundary of the turbulent wake.

As in the case of a laminar wake, we represent the velocity of the fluid in the form $\mathbf{u}+\mathbf{v}$, where $v \ll u$. The quantity y is the width of the turbulent wake at a distance $x \gg R$ from the body. We then have the following equation for the boundary of a turbulent wake:

$$\frac{dy}{dx} \sim \frac{v}{u}. \tag{8.16}$$

Indeed, the velocity along the stream direction is equal to u, since the correction v can be neglected. In the transverse direction y the velocity of the fluid in the turbulent wake is of order v. In contrast to the case of a laminar wake, all of the components of the velocity v are of the identical order of magnitude because of the vortex nature of turbulent motion.

The expression (7.34) for the drag force on a body also holds in this case, where the x component of the velocity v_x can be replaced by v:

$$F \sim \rho v u y^2. \tag{8.17}$$

Substituting (8.17) into (8.16), we can rewrite (8.16) in the form

$$\frac{dy}{dx} \sim \frac{F}{\rho u^2 y^2}. \tag{8.18}$$

Replacing the derivative dy/dx on the left-hand side of (8.18) by the ratio y/x, we find the width y of the turbulent wake as a function of the distance x from the body:

$$y \sim \left(\frac{Fx}{\rho u^2}\right)^{1/3}. \tag{8.19}$$

The drag force F on the body is not given by Stokes's law (7.20), which holds only for small Reynolds numbers $\mathrm{Re} \ll 1$. When $\mathrm{Re} \gg 1$ the entire resistance to the flow is kinematic. The kinematic pressure is given by (8.4) and therefore the drag force on a body with linear dimension R is of order

$$F \sim \rho u^2 R^2. \tag{8.20}$$

Substituting (8.20) into (8.19), we find for the width of the wake

$$y \sim (x/R)^{1/3} R \gg R. \tag{8.21}$$

From this relation we find

$$(y/x) \sim (R/x)^{2/3} \ll 1,$$

which was already used in the derivation of (8.16).

We note that the width of a turbulent wake (8.21) is much smaller than the width of a laminar wake (7.28), for the same value of x. Therefore as the stream velocity u increases the width of the wake decreases until the flow becomes turbulent, in which case the width no longer depends on the stream velocity u [see (8.21)].

We estimate the typical deviation v of the velocity in a turbulent wake from the stream velocity u. From (8.16) and (8.21) we find

$$v \sim \frac{y}{x} u \sim \left(\frac{R}{x}\right)^{2/3} u \ll u. \tag{8.22}$$

With the help of this relation we can estimate the Reynolds number Re_x at a distance x from the body in the turbulent wake. If we transform to a reference frame in which the body moves, while the fluid is at rest at infinity, the velocity u of the body obviously drops out of the problem and the Reynolds number is determined by the velocity v. The width y of the turbulent wake obviously serves as the characteristic length. Therefore we find $\mathrm{Re}_x = vy/\nu$. Using (8.21) and (8.22), we obtain

$$\mathrm{Re}_x \sim \left(\frac{R}{x}\right)^{1/3} \mathrm{Re}. \tag{8.23}$$

We see from (8.23) that the Reynolds number Re_x decreases as x increases. When $\mathrm{Re}_x \sim 1$ the turbulent wake transforms into a laminar wake. This takes place at the distance

$$x \sim (\mathrm{Re})^3 R. \tag{8.24}$$

In the transition from a turbulent wake to a laminar wake, the rate of expansion of the wake with distance x from the body increases.

The width of a turbulent wake far behind a long cylinder of radius R as a function of the distance $x \gg R$ from the body can be estimated using the same method. We assume that the stream velocity u is perpendicular to the axis of the cylinder and that the Reynolds number $\mathrm{Re} = uR/\nu \gg 1$.

The general equation (8.16) for the boundary of the turbulent wake remains valid in this case. However, the drag force F differs from (8.17): in place of the area y^2 we use the width y of the turbulent wake and then the expression

$$F \sim \rho v u y$$

represents the drag force per unit length of the cylinder. In place of (8.18) we obtain

$$\frac{dy}{dx} \sim \frac{F}{\rho u^2 y}. \tag{8.25}$$

Again, setting $dy/dx \sim y/x$ and using the fact that $F \sim \rho u^2 R$ as in (8.20), we find from (8.25)

$$y \sim \left(\frac{x}{R}\right)^{1/2} R \gg R. \tag{8.26}$$

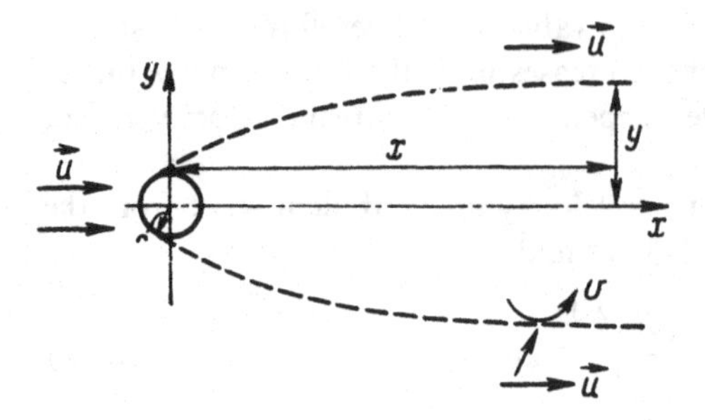

FIG. 13. Turbulent wake behind a cylinder of radius R in a fluid with stream velocity u. The boundary of the wake is shown by the dashed lines; v is the velocity of vortex motion inside the wake.

The boundary of the turbulent wake in this case is a parabola (Fig. 13).

Using (8.16), we can estimate the velocity v in the turbulent wake behind the cylinder:

$$v \sim \left(\frac{R}{x}\right)^{1/2} u \ll u. \qquad (8.27)$$

It follows from (8.26) and (8.27) that the Reynolds number $\mathrm{Re}_x = vy/\nu$ is independent of the distance x and is roughly equal to the Reynolds number $\mathrm{Re} = uR/\nu$ near the body. Therefore, in contrast to the previous problem, here there is no transition from a turbulent wake to a laminar wake.

Problem 1. Show that the longitudinal velocity v_x outside the turbulent wake falls off with distance x from the body as x^{-2}, and therefore decreases with distance much more rapidly than inside the turbulent wake [from (8.22) the longitudinal velocity decreases with x inside the wake as $x^{-2/3}$].

Problem 2. Show that in the case of a fluid flowing past a long cylinder in the direction perpendicular to the cylinder axis the velocity v of the fluid outside the turbulent wake falls off with distance x from the cylinder as x^{-1}, and therefore decreases with distance much more rapidly than inside the turbulent wake [from (8.27) the velocity inside the wake decreases with x as $x^{-1/2}$].

8.3. Relaxation of turbulent flow

We consider how a turbulent fluid returns to a state of rest after the motion of the body causing the turbulence stops. Let a body with linear dimension R move in a fluid with velocity u and then stop. The Reynolds number $\mathrm{Re} = uR/\nu$ is assumed to be large, i.e., $\mathrm{Re} \gg 1$.

We consider a volume of the fluid containing a large number of vortices randomly distributed with respect to one another. We transform to a reference frame in which the volume as a whole is at rest. For example, in the turbulent wake problem this is the reference frame in which the body moves, and the fluid at infinity is at rest.

The angular momentum vectors of the vortices are added together to obtain the total angular momentum of the volume

$$\mathbf{L}=\sum_i \mathbf{L}_i. \tag{8.28}$$

When the source of the turbulent motion is turned off the total angular momentum must remain constant, since the volume of fluid under consideration is effectively an isolated system (neglecting currents through the surface surrounding the volume). Therefore we assume that the volume of fluid in which the turbulent motion relaxes is effectively bounded by walls.

Suppose there are N vortices of a certain size l inside the volume. The total angular momentum (8.28) will then be attributed to vortices of this size, where the index i enumerates the vortices. Here, we assume that momentum and energy exchange between vortices of different sizes is slow. This exchange is determined by dissipative processes, and was discussed in the introduction to the present chapter.

From the conservation of the total angular momentum (8.28) we will also obtain an estimate of the time required for the turbulent fluid to reach a state of rest. We first estimate the angular momentum for each vortex:

$$L_i \sim \rho V_i l v_l. \tag{8.29}$$

Here, l is the linear dimension of the vortex, v_l is its typical velocity, ρ is the density of the fluid, and V_i is the volume of fluid occupied by the vortex (ρV_i is the mass of the vortex). Because $V_i \sim l^3$, we obtain from (8.29)

$$L_i \sim \rho v_l l^4. \tag{8.30}$$

The square of the total angular momentum (8.28) of the fluid volume is the sum of the squares of the angular momenta of the individual vortices L_i^2 and their pairwise products $L_i L_j$. The pairwise products vanish when averaged over time because the motion of different vortices is uncorrelated. The assumption that the vortices are independent is a hypothesis. Accepting this hypothesis, we find that

$$L^2 \sim N L_i^2. \tag{8.31}$$

The number of vortices N in the fluid volume is roughly $N \sim V/V_i$, where V is the total volume of the region of fluid under consideration. We have assumed that at a given instant of time only vortices of some size l exist; the size changes in time (as we will see, it increases) and therefore the velocity v_l also changes (as we will see, it decreases).

Therefore we find from (8.31)

$$L^2 \sim V L_i^2 / l^3. \tag{8.32}$$

Substituting (8.30) into (8.32), we find

$$L^2 \sim V \rho^2 v_l^2 l^5. \tag{8.33}$$

From the conservation of this quantity we obtain a relation between l and v_l: $v_l \sim l^{-5/2}$. We must have $v_l \sim u$ when $l \sim R$; therefore,

$$v_l \sim (R/l)^{5/2} u \tag{8.34}$$

(the **Loĭtstyanskiĭ law**).

Note the difference between the estimates (8.5) and (8.34) for the velocity v_l. The estimate (8.5) holds for the steady case of fully developed turbulence maintained by an external source, such as a moving body. The estimate (8.34) describes the connection between the characteristics of the vortex after the external source is turned off and the motion relaxes toward the state of rest.

We determine the time dependence of the quantities v_l and l in the process of this relaxation. From dimensional considerations we can write

$$l \sim v_l t \tag{8.35}$$

since the viscosity ν cannot appear in the problem for large Reynolds numbers $\text{Re} \gg 1$ because the fluid is ideal (see the introduction to the present chapter). In the case of the largest vortices, the estimate (8.35) is consistent with (8.3), as it must be.

From (8.34) and (8.35) we obtain

$$v_l \sim [R/(ut)]^{5/7} u, \quad l \sim (ut/R)^{2/7} R. \tag{8.36}$$

Hence *after the motion of the body ceases the velocity of a turbulent vortex decreases in time and its size increases.* It easily follows from (8.36) that the Reynolds number $\text{Re}_t = v_l l/\nu$ as a function of time t decreases as $t^{-3/7}$. Hence turbulent motion gradually diminishes.

Setting the Reynolds number Re_t equal to unity, we estimate the damping time of a strongly agitated fluid in a closed volume

$$\tau \sim (\text{Re})^{7/3} \frac{R}{u} \gg \frac{R}{u}. \tag{8.37}$$

We compare this time with the relaxation time (7.25) of laminar flow for the same initial linear dimension R. The ratio of (8.37) to the time R^2/ν given by (7.25) is of order $(\text{Re})^{4/3} \gg 1$. Therefore *the turbulent relaxation time is very large in comparison with the laminar relaxation time.*

In the final stage of turbulent relaxation the estimate (8.35) is incorrect, because the inertial forces become small in comparison with the viscous forces. Using (7.25), we find that in place of (8.35) we have $l \sim (\nu t)^{1/2}$. Substituting this estimate into (8.34), we find the velocity v_l as a function of time in the final stage of turbulence:

$$v_l \sim \left(\frac{R \cdot \text{Re}}{ut} \right)^{5/4} u \tag{8.38}$$

(**Millionshchikov's law**).

Using the velocity (8.38) and the linear dimension of the flow $l \sim (\nu t)^{1/2}$, to again estimate the Reynolds number $\text{Re}_t = v_l l/\nu$, we again obtain (8.37) for the relaxation time τ of turbulent flow (defined to be the time such that $\text{Re}_\tau \sim 1$).

Problem 1. Explain why the estimate (8.36) does not hold, while (8.38) remains correct if the turbulent motion damps out in an infinite fluid.

8.4. Onset of turbulence

In this section we consider a mechanism which qualitatively explains the onset of turbulent motion of a fluid with increasing Reynolds number. This mechanism was discussed by Feigenbaum.

Qualitatively, we can write the Navier–Stokes equation (7.3) in the form

$$\frac{\partial v}{\partial t}=av^2+b+cv. \tag{8.39}$$

Suppose the motion of the fluid is periodic with period T. Then, the left-hand side of (8.39) can be rewritten in the form

$$\frac{\partial v}{\partial t}\approx[v(t+T)-v(t)]/T. \tag{8.40}$$

Substituting (8.40) into (8.39), we can rewrite the Navier–Stokes equation for the velocity v in the form of a difference equation

$$v(t+T)=\alpha v(t)+\beta v^2(t)+\gamma. \tag{8.41}$$

Here, α, β, and γ are constants. The linear term $\alpha v(t)$ in (8.41) can be omitted with no loss of generality by transforming to a reference frame moving with respect to the original frame by a certain constant velocity V, i.e., $v(t)\rightarrow v(t)+V$. Then (8.41) takes the form

$$v(t+T)=u+\lambda v^2(t)/u. \tag{8.42}$$

Here, u is a certain characteristic velocity of the fluid, such as the velocity of a body moving in the fluid. The quantity λ is a dimensionless constant, which qualitatively can be identified with the Reynolds number Re, since the ratio of the nonlinear term in the Navier–Stokes equation to the viscous term is of the order of the Reynolds number [see the discussion after (7.4)].

Therefore we obtain from (8.42) a difference equation which qualitatively retains all of the main features of the original Navier–Stokes equation

$$v(t+T)=u-\mathrm{Re}\cdot v^2(t)/u. \tag{8.43}$$

The minus sign in front of the quadratic term in this equation is chosen because otherwise the velocity v would increase without bound as the motion repeated in time with period T ($u>0$).

A strictly periodic solution of (8.43) implies that $v_1(t+T)=v_1(t)$. Therefore we obtain from (8.43) the quadratic equation

$$v_1=u-\mathrm{Re}\cdot v_1^2/u, \tag{8.44}$$

whose solution is

$$v_1=\frac{\sqrt{1+4\,\mathrm{Re}}-1}{2\,\mathrm{Re}}\,u. \tag{8.45}$$

Here, we have chosen the positive root of the equation.

We now consider a small perturbation on this periodic solution: $v = v_1 + \Delta v$. Substituting this solution into (8.43) and keeping terms up to second order in the perturbation Δv, we find

$$\Delta v(t+T) = -2\,\mathrm{Re}\cdot v_1 \Delta v(t)/u. \tag{8.46}$$

If

$$2\,\mathrm{Re}\cdot v_1 < u, \tag{8.47}$$

then according to (8.46) $|\Delta v(t+T)| < |\Delta v(t)|$ and the perturbation decreases in time. Therefore periodic motion with the velocity (8.45) is stable. In the opposite case $2\,\mathrm{Re}\cdot v_1 > u$ the motion is unstable because the perturbation grows rapidly in time. The loss of stability occurs for values $\mathrm{Re} = \mathrm{Re}_1$, $v_1 = \langle v_1 \rangle$, where

$$2\,\mathrm{Re}_1 \langle v_1 \rangle = u. \tag{8.48}$$

Eliminating $\langle v_1 \rangle$ between the two equations (8.45) and (8.48), we find

$$\mathrm{Re}_1 = 3/4. \tag{8.49}$$

Hence *periodic motion with velocity v_1 is stable for Re < Re_1 and is unstable for Re > Re_1.*

According to (8.46), we have $\Delta v(t+T) = -\Delta v(t)$ on the stability boundary and therefore $\Delta v(t+2T) = \Delta v(t)$. We see that motion with the velocity $\langle v_1 \rangle + \Delta v$ is periodic with period $2T$, i.e., with double the period of the original periodic motion. This assertion is correct not only when $\mathrm{Re} = \mathrm{Re}_1$, but for any Re. However, in the general case Δv is not an arbitrarily small perturbation, as above, but has a certain finite value corresponding to the velocity $v = \langle v_1 \rangle + \Delta v$ of the flow with period $2T$.

We find the condition for loss of stability of the periodic flow with period $2T$. We iterate (8.43) through another period

$$v(t+2T) = u(1-\mathrm{Re}) + 2\,\mathrm{Re}^2 \cdot v^2(t)/u - \mathrm{Re}^3 \cdot v^4(t)/u^3. \tag{8.50}$$

The last term on the right-hand side of (8.50) can be neglected [it can be shown that $v < u$ and therefore $(v/u)^4$ is small]. We introduce the notation

$$v(t) = \tilde{v}(t)(1-\mathrm{Re}), \quad \lambda = 2\,\mathrm{Re}^2(\mathrm{Re}-1). \tag{8.51}$$

Then (8.50) takes the form

$$\tilde{v}(t+2T) = u - \lambda \tilde{v}^2(t)/u. \tag{8.52}$$

We see that this equation has the same form as (8.43) if we replace v by $\tilde{v}$ and Re by λ. Repeating the stability analysis given above, we obtain from (8.49) that loss of stability of the periodic motion with period $2T$ occurs when $\lambda = \lambda_2 = 3/4$. From (8.51) the corresponding Reynolds number is $\mathrm{Re}_2 = 1.23$.

Therefore *periodic motion with the doubled period $2T$ is stable when Re < $\mathrm{Re}_2 = 1.23$.*

Performing another iteration of (8.43), it can be shown that loss of stability of periodic motion with period $4T$ occurs when $\mu = 3/4$, which is the analog of the condition (8.49), where $\mu = 2\lambda^2(\lambda - 1)$ [from (8.51)]. Therefore $\lambda = \lambda_3 = 1.23$ and we find from (8.51) that $\mathrm{Re}_3 = 1.34$.

Repeating this procedure an infinite number of times, we obtain that the total loss of stability of flow with all periods takes place for the critical Reynolds number $\mathrm{Re}_{cr} = \mathrm{Re}_\infty$. According to (8.51) this number is found from the equation

$$\mathrm{Re}_{cr} = 2\,\mathrm{Re}_{cr}^2(\mathrm{Re}_{cr} - 1). \tag{8.53}$$

Hence we easily obtain

$$\mathrm{Re}_{cr} = (1+\sqrt{3})/2 \approx 1.37.$$

The region $\mathrm{Re} > \mathrm{Re}_{cr}$ is the region we call turbulent flow. In this region all vortex motion is unstable and it is rapidly destroyed a short time after its inception.

Note the rapid convergence of the sequence of Reynolds number Re_n with increasing n corresponding to loss of stability of periodic motion with period $2^{n-1}T$.

Period doubling of the motion is one of the ways that turbulence originates.

Problem 1. Show that the second root of the quadratic equation (8.44) corresponds to unstable periodic motion for all Reynolds numbers.

Problem 2. Using the approach of this section, consider the breakdown of periodic motion when the velocity perturbation satisfies $\Delta v(t+T) = \Delta v(t)$, and not $-\Delta v(t)$, as above. Show that when the critical Reynolds number is exceeded, the trajectories of fluid particles after a long time are almost periodic, i.e., the flow is laminar. However, after a certain time the trajectory becomes unstable and then the process repeats. Obtain the following estimate for the time interval τ in which the motion is laminar

$$\tau \sim (\mathrm{Re} - \mathrm{Re}_{cr})^{-1/2},$$

where $\mathrm{Re} - \mathrm{Re}_{cr}$ is the supercriticality of the flow.

Chapter 9

Boundary layers

A *boundary layer* is a region near the surface of a body in a moving fluid whose thickness is small in comparison with the linear dimension R of the body. It is characterized by a large velocity gradient in the direction perpendicular to the surface at the point in question. The gradient is large at large Reynolds number because the fluid velocity is equal to zero at the surface of the body and it is very large a small distance away, where it approaches the stream velocity of the fluid.

Boundary layers can be laminar or turbulent. A laminar boundary layer transforms into laminar flow at a certain distance from the surface of the body. The viscosity of the fluid cannot be neglected inside the laminar boundary layer, even though the Reynolds number is large. This is because inside the boundary layer the important distance in the problem is the thickness of the layer and therefore the local Reynolds number is small. For a body moving in a fluid a laminar boundary layer is formed in the region in the front of the body. A turbulent boundary layer is formed behind the body, and it borders on the region of turbulent motion of the fluid.

As noted in the preceding chapter, laminar flow transforms into turbulent flow at a certain critical Reynolds number much larger than unity. The actual value of the critical Reynolds number depends on the configuration of the bodies in the problem. In addition, for a given configuration the critical Reynolds number depends strongly on the amplitudes of initial perturbations in the incoming fluid stream: the larger these amplitudes, the larger the critical Reynolds number. These same factors also affect the structure of the boundary layers adjacent to the laminar and turbulent flow fields.

9.1. Laminar boundary layer

We estimate the thickness of the laminar boundary layer near the surface of a body with linear dimension R in a fluid with velocity u. The Reynolds number Re $= uR/\nu \gg 1$.

We consider steady laminar flow. We take the velocity u along the x axis and define the y axis to be perpendicular to the surface of the body. We further assume that the parameters of the problem are independent of the z coordinate (Fig. 14). The variation with x is characterized by the linear dimension R of the body. Since the parameters of the problem are independent of z, the incompressibility equation of the fluid (1.1) takes the form

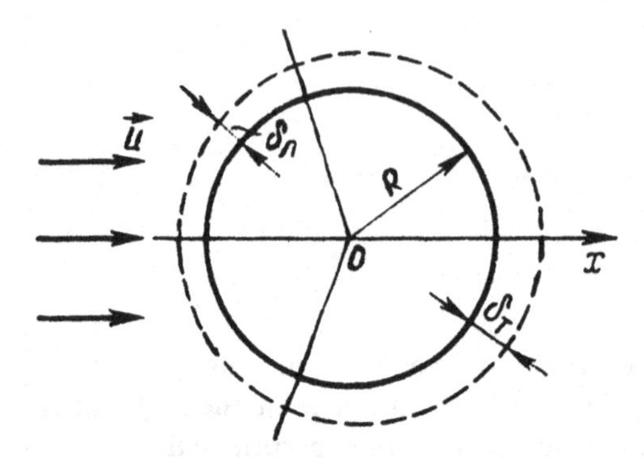

FIG. 14. Laminar (thickness δ_l) and turbulent (thickness δ_t) boundary layers of a body of linear dimension R in a fluid moving with a large velocity u.

$$\frac{\partial v_x}{\partial x}+\frac{\partial v_y}{\partial y}=0. \tag{9.1}$$

Roughly,

$$\frac{v_x}{R}\sim\frac{v_y}{\delta}. \tag{9.2}$$

Here, δ is the thickness of the boundary layer along y. The velocity v_x is of the same order of magnitude as the stream velocity u outside the laminar boundary layer. Therefore

$$v_y\sim\frac{\delta}{R}u\ll u. \tag{9.3}$$

This inequality will be proven below, and it will be seen that the thickness δ of the boundary layer is small in comparison with the size of the body R.

We turn now to the Navier–Stokes equation (7.4) for steady flow

$$v_x\frac{\partial \mathbf{v}}{\partial x}+v_y\frac{\partial \mathbf{v}}{\partial y}=-\frac{1}{\rho}\nabla P+\nu\left(\frac{\partial^2\mathbf{v}}{\partial x^2}+\frac{\partial^2\mathbf{v}}{\partial y^2}\right). \tag{9.4}$$

Taking the y component of this equation, approximating each of the terms, and assuming that $\delta\ll R$, we find

$$u\frac{v_y}{R}\sim\frac{P}{\rho\delta}\sim\nu\frac{v_y}{\delta^2}. \tag{9.5}$$

Hence this equation determines the thickness of the boundary layer δ:

$$\delta\sim\left(\frac{\nu R}{u}\right)^{1/2}\sim\frac{R}{(\mathrm{Re})^{1/2}}\ll R. \tag{9.6}$$

Here, the Reynolds number is $\mathrm{Re}=uR/\nu\gg 1$.

Substituting (9.6) into (9.3), we obtain an estimate for the velocity component v_y:

$$v_y\sim\frac{u}{(\mathrm{Re})^{1/2}}\ll u. \tag{9.7}$$

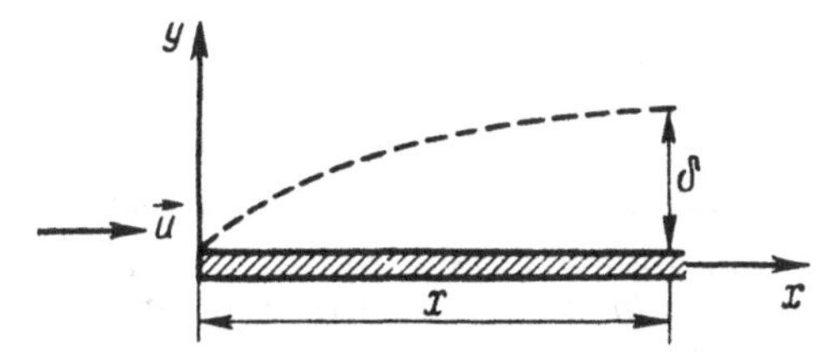

FIG. 15. Laminar boundary layer of a half plane in a moving fluid; δ is the thickness of the boundary layer.

We can then estimate the pressure P from (9.5):

$$P \sim \frac{\rho u^2}{\text{Re}} \ll \rho u^2. \tag{9.8}$$

Taking the x component of the vector equation (9.4), we obtain

$$\frac{u^2}{R} \sim \nu \frac{u}{\delta^2}, \tag{9.9}$$

and hence we again obtain (9.6) for the thickness of the boundary layer. The term involving the pressure in (9.4) can be neglected, since it is of order $P/(\rho R)$, and the ratio of this term to the first term in (9.9) is of order $P/(\rho u^2) \sim \text{Re}^{-1} \ll 1$.

As noted above, a laminar boundary layer is formed in the region in front of the body. It begins at the leading point of the body with respect to the incoming stream (Fig. 14) and spreads along the surface of the body to a distance determined by the shape of the body. At this point the laminar boundary layer transforms into a turbulent boundary layer (see Sec. 9.2). For a circular cylinder the experimental value of the angle between the leading point on the body and the point where the flow becomes turbulent is about 75°. The boundary between the laminar and turbulent boundary layers is shown by the dash-dotted line in Fig. 14.

We estimate the thickness of the laminar boundary layer of a plate in the form of a half plane (Fig. 15) in a fluid moving with velocity u parallel to the plane of the plate and perpendicular to its edge. As above, the Reynolds number is assumed to be large. The distance x from the plate edge in the direction of motion of the fluid plays the role of the characteristic linear dimension in the Reynolds number.

Let the x axis be along the stream velocity u of the fluid and let y be perpendicular to the plane of the plate (Fig. 15). The z axis is along the plate edge and we assume that $x = y = 0$ at the edge. We will see below that the thickness of the boundary layer increases with increasing distance x from the edge of the plate.

The discussion of the preceding problem remains valid here, except that the linear dimension R of the body in the x direction is replaced by the x coordinate of the point on the plate under consideration. We then obtain from (9.6)

$$\delta \sim \left(\frac{\nu x}{u}\right)^{1/2}. \tag{9.10}$$

Hence the thickness δ of the layer varies with x as $x^{1/2}$. The boundary of the layer is shown by the dashed line in Fig. 15.

We estimate the frictional force per unit area acting on the plate. Since the flow is laminar we can use (7.1):

$$F_x = \eta \frac{dv_x}{dy} \sim \eta \frac{u}{\delta} \sim \left(\frac{\rho \eta u^3}{x}\right)^{1/2}. \tag{9.11}$$

Here, we have used (9.10) for the thickness δ. We see that the *frictional force falls off with increasing distance x from the plate edge.* If the length of the plate along x is l and the width along z is b, then the total frictional force acting on the plate can be obtained by multiplying (9.11) by the area bl of the plate, where we set $x \sim l$ in (9.11). We then obtain

$$F \sim F_x bl \sim b(\rho \eta u^3 l)^{1/2}. \tag{9.12}$$

The force acts on the lower side of the half plane. We note that the proportionality constant in (9.12) is equal to 1.33 (when the fluid flows past both sides of the half plane). Then (9.12) is called **Blasius's law**. Recall that this formula holds when $\mathrm{Re} = ul/\nu \gg 1$. We see from (9.12) that the frictional force is proportional to $u^{3/2}$. In the case of small Reynolds number $\mathrm{Re} \ll 1$ it follows from (7.1) that the frictional force acting on a unit area of the plate is $F_x \sim \eta u/l$ and therefore the total drag force on the plate $F \sim F_x lb \sim \eta ub$ is proportional to the first power of the stream velocity u. Therefore as the velocity increases, the drag force on the plate first increases linearly with u and then more rapidly (as $u^{3/2}$).

Next, we estimate the thickness of the boundary layer for a disk of radius R rotating uniformly with large angular velocity ω about its axis in a fluid. The linear velocity u at a distance r from the axis of the disk is ωr. Substituting this value of u in (9.10), and also putting r in place of x, we obtain an estimate for the thickness of the boundary layer on the surface of the rotating disk

$$\delta \sim \left(\frac{\nu}{\omega}\right)^{1/2}. \tag{9.13}$$

We see that the thickness δ does not depend on the coordinate r and so the laminar boundary layer has the same thickness over the entire surface of the disk, as in the case of a finite body in a moving fluid [compare (9.6)]. The result (9.13) holds when $\mathrm{Re} = \omega R^2/\nu \gg 1$.

Finally, we consider the formation of a laminar boundary layer when a fluid flows into a pipe of linear dimension R with a large velocity u such that the Reynolds number $\mathrm{Re} = uR/\nu \gg 1$. Near the end of the pipe the fluid velocity is approximately constant across a cross section of the pipe and is equal to u almost everywhere except for a thin boundary layer near the inner surface of the pipe, where it rapidly decreases to zero. Let x be the distance along the pipe axis from the pipe opening. According to (9.10), the thickness δ of the boundary layer increases with increasing distance x.

At a certain value of x the thickness of the boundary layer becomes equal to the pipe radius, i.e., $\delta \sim R$. We determine this value of x from (9.10):

$$x \sim \frac{uR^2}{\nu} \sim \mathrm{Re} \cdot R \gg R. \tag{9.14}$$

Beginning with distances of this order we have steady viscous Poiseuille flow with a parabolic variation of velocity in the radial direction (see Sec. 7.1). Because

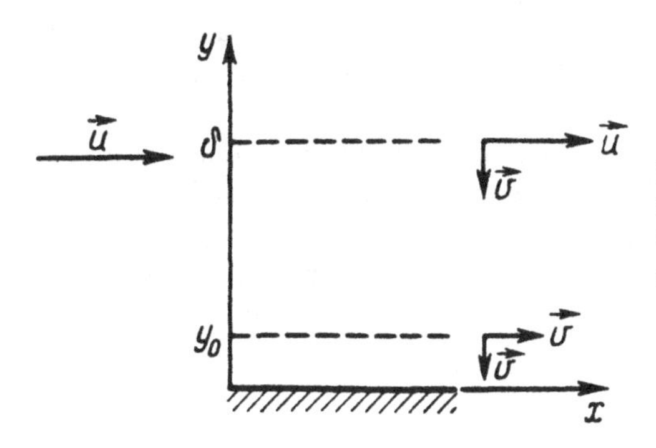

FIG. 16. Turbulent boundary layer: u is the stream velocity of the fluid, δ is the thickness of the boundary layer, and y_0 is the thickness of the viscous sublayer.

Re$\gg$1, the distance (9.14) is large in comparison with the pipe radius R. We note that for a pipe of circular cross section of radius R the numerical factor in (9.14) is equal to 0.046.

The above results for flow in a pipe are valid for Reynolds numbers $1 \ll \mathrm{Re} < \mathrm{Re}_{cr}$, where $\mathrm{Re}_{cr} \gg 1$ determines the transition from laminar flow to turbulent flow. Hence the results of Sec. 7.1 for Poiseuille flow are also correct under these conditions, and not just for Re$\ll$1.

Problem 1. Show that the thickness of the laminar boundary layer at the leading point of a body in a moving fluid (Fig. 14) is also given by the estimate (9.6).

Problem 2. A fluid moves in a slit of expanding width in the direction of expansion of the slit. The Reynolds number $\mathrm{Re}_x = v(x)x/\nu \gg 1$, where $v(x)$ is the velocity of the fluid at distance from the edge of the slit. The flow rate of fluid per unit length of the slit is equal to Q. Show that the thickness of the laminar boundary layer along the edges of the slit increases with x according to the equation

$$\delta \sim x(\eta/Q)^{1/2}. \tag{9.15}$$

9.2. Turbulent boundary layer

As noted above, a turbulent boundary layer is formed in the region behind the body. We estimate the thickness of the turbulent boundary layer near the surface of a body with linear dimension R in a fluid with stream velocity u. The Reynolds number $\mathrm{Re} = uR/\nu$ is assumed to be large in comparison with unity and larger than the critical Reynolds number.

We consider a small area on the surface of the body and let x be along the surface parallel to the velocity vector u in this region (Fig. 16). Let δ be the thickness of the boundary layer along y, where the y axis is perpendicular to the surface of the body in the region under consideration. We assume that the parameters of the problem are independent of the z coordinate.

Consider a unit volume of fluid at a distance y from the surface of the body. According to (8.2), the power dissipated inside this volume is of order

$$j \sim \rho \frac{v^3}{y}. \tag{9.16}$$

Here, ρ is the density of the fluid and v is the velocity of vortex motion in the turbulent boundary layer. Earlier we assumed that it was of the same order of magnitude as the stream velocity u. We will see that this is actually true only to within a logarithmic factor and $v \ll u$.

It follows from (9.16) that as y increases the rate of energy dissipation decreases. The rate of energy dissipation is much larger than in the case of laminar flow, where we have $j \sim \eta v^2/y$, according to (7.46).

The dissipated power per unit surface area of the body $\dot{E}$ is obtained from (9.16) by integrating with respect to y

$$\dot{E} \sim \int_{y_0}^{\delta} j\,dy \sim \rho v^3 \ln \frac{\delta}{y_0}. \tag{9.17}$$

Here, the lower limit corresponds to a Reynolds number of order unity, since (9.16) is not correct for smaller values of y and the contribution of the viscous sublayer ($y < y_0$) to the energy dissipation is small in comparison with (9.17). Therefore y_0 is found from the condition

$$\frac{v y_0}{\nu} \sim 1. \tag{9.18}$$

As y increases from y_0 to a value of order δ, the horizontal component of the fluid velocity v_x in the boundary layer increases from v to u, while the vertical component v_y remains of order v. In (9.16) we use the vertical component $v_y \sim v$ and not the horizontal component v_x, since the latter can be changed by transforming to a coordinate system moving with respect to the original system.

The rate of energy dissipation into heat (9.17) is roughly equal to the work done by the frictional forces per unit area in the region of the surface under consideration. A part of the energy can be transformed into the kinetic energy of the perturbed motion of the fluid. But for turbulent flow all of these energies are of the same order of magnitude because energy dissipation in a turbulent fluid results from the transfer of kinetic energy between different vortices (see the introduction to Chap. 8).

Let A be the work done by the frictional forces per unit surface area of the body and per unit time (i.e., the power). Obviously,

$$A \sim F_x u. \tag{9.19}$$

Here, u is the velocity of the body in the fluid (or the velocity of the fluid impinging on the body) and F_x is the drag force per unit area at the region of the surface under consideration. In the case of turbulent flow (8.4) gives

$$F_x \sim \rho v^2. \tag{9.20}$$

Here, we have substituted the velocity component $v_y \sim v$ perpendicular to the surface of the body; this is the velocity component responsible for the kinematic pressure on the body.

Equating (9.17) and (9.19), we obtain an estimate for the velocity v:

$$v \sim \frac{u}{\ln(\delta/y_0)}. \tag{9.21}$$

Replacing δ by R and y_0 by ν/u [see (9.18)] in the argument of the logarithm, we obtain from (9.21)

$$v \sim \frac{u}{\ln(\text{Re})} \ll u, \tag{9.22}$$

which confirms that the velocity v of vortex motion of the fluid in the boundary layer is small in comparison with the stream velocity u of the fluid. Therefore as y increases from y_0 to δ the horizontal component of the velocity v_x increases from v to $v \ln(\delta/y_0)$:

$$v_x \sim v \ln\left(\frac{y}{y_0}\right) \tag{9.23}$$

(the so-called *logarithmic velocity profile*).

Substituting (9.22) into (9.18), we obtain an estimate for y_0:

$$y_0 \sim \frac{\nu}{u} \ln(\text{Re}) \sim \frac{\ln(\text{Re})}{\text{Re}} R \ll R. \tag{9.24}$$

We estimate the thickness of the boundary layer δ. Let $y \sim \delta$, which corresponds to the edge of the turbulent boundary layer. As we have seen in Chap. 8, at the edge fluid can only flow into the turbulent boundary layer, and cannot flow out. In analogy with (8.16), the condition that the streamlines of the fluid on the side of the turbulent boundary layer must be parallel to the edge of the layer can be written in the form

$$\frac{dy}{dx} \sim \frac{v_y}{v_x}, \quad \text{or} \quad \frac{\delta}{R} \sim \frac{v}{u}. \tag{9.25}$$

Substituting (9.22) into (9.25), we finally obtain

$$\delta \sim \frac{R}{\ln(\text{Re})} \ll R. \tag{9.26}$$

Comparing (9.24) and (9.26), we see that $y_0 \ll \delta$, as expected.

If we have a half plane in a moving fluid, rather than a body of linear dimension R, all of the above relations remain valid except that R is replaced by the distance x from the edge of the plate to the point in question along the plane of the plate (recall that z is the direction of the plate edge and y is the direction perpendicular to the plane of the plate). In particular, instead of (9.26) we obtain for the thickness of the boundary layer δ:

$$\delta \sim \frac{x}{\ln(ux/\nu)} \ll x. \tag{9.27}$$

Instead of (9.22) we have for the velocity v of vortex motion in the turbulent boundary layer

$$v \sim \frac{u}{\ln(ux/\nu)} \ll u. \tag{9.28}$$

Comparing (9.6) and (9.26), we see that a turbulent boundary layer is much thicker than a laminar one. Hence *the boundary layer behind the body is thicker than the one in front of it.*

In the case of a half plane the Reynolds number $\mathrm{Re} = ux/\nu$ increases with the distance x from the edge of the half plane and at a certain critical Reynolds number $\mathrm{Re}_{cr} \gg 1$ laminar flow transforms into turbulent flow, accompanied by a transition of the laminar boundary layer into a turbulent boundary layer. Then the thickness of the boundary layer is given by (9.27) instead of (9.10), i.e., it is much larger.

We estimate the velocity u of turbulent flow through a pipe of radius R and length l ($l \gg R$) with a pressure difference δP across the two ends of the pipe. The Reynolds number $\mathrm{Re} = uR/\nu \gg 1$.

This problem complements the problem considered in Sec. 7.1, where the Reynolds number was assumed to be small. We consider the region near the inner surface of the pipe where the turbulent boundary layer is formed. Then the surface of the pipe at a given location can be treated approximately as a plane. From (9.27), where x is the distance from the beginning of the pipe along the pipe axis, we find that the thickness of the boundary layer increases as the fluid moves along the axis of the pipe. When $x \sim R$ it reaches a value equal to the radius of the pipe and obviously cannot increase any further. Therefore we obtain the velocity v of vortex motion in the pipe from (9.28)

$$v \sim \frac{u}{\ln(uR/\nu)}. \tag{9.29}$$

Here, u is the stream velocity of the fluid through the pipe.

The flow rate Q is defined in terms of u in the usual way: $Q \sim \rho u R^2$. Substituting (9.29) into this formula, we obtain

$$Q \sim \rho v R^2 \ln(\mathrm{Re}). \tag{9.30}$$

We relate the quantities v and δP, noting that a pressure difference δP leads to a force of order $\delta P \cdot R^2$ on the fluid inside the pipe. In the case of steady flow this force must be equal to the frictional force of the fluid against the pipe. The frictional force F_x acting on a unit surface area of the pipe is given by (9.20). The total frictional force on the whole pipe is equal to the product of F_x and the surface area of the pipe $\sim Rl$. Therefore

$$\rho v^2 R l \sim \delta P R^2,$$

and hence we obtain for the velocity v

$$v \sim \left(\frac{R\delta P}{\rho l}\right)^{1/2}. \tag{9.31}$$

Substituting (9.31) into (9.29), we obtain the stream velocity u:

$$u \sim \left(\frac{R\delta P}{\rho l}\right)^{1/2} \ln\left(\frac{R^3 \delta P}{\nu^2 \rho l}\right). \tag{9.32}$$

This result involves the combination $\delta P/l$, as it must, since this combination is the pressure drop per unit length of the pipe. It is also evident from (9.32) that the stream velocity basically depends on the pipe radius as $u \sim R^{1/2}$, i.e., much more weakly than in the case of viscous Poiseuille flow at small Reynolds number, where $u \sim R^2$ [see (7.6)].

Substituting (9.31) into (9.30), we obtain the flow rate Q

$$Q \sim \left(\frac{\rho R^5 \delta P}{l}\right)^{1/2} \ln\left(\frac{R^3 \delta P}{\nu^2 \rho l}\right). \tag{9.33}$$

We compare the energy dissipation in a pipe in the case of laminar (Poiseuille) flow and in the case of turbulent flow. The power dissipated per unit volume of the pipe in the case of laminar flow is given by (7.46) and has the form

$$j \sim \rho\nu \frac{u^2}{R^2}. \tag{9.34}$$

Multiplying this expression by the volume of fluid inside the pipe ($\sim R^2 l$), we obtain the dissipated power E in the entire pipe

$$\dot{E} \sim \rho\nu u^2 l. \tag{9.35}$$

The dissipated power in turbulent flow per unit surface area of the pipe is given by (9.17) and has the form [using (9.24), we substitute R in place of δ and ν/u in place of y_0]:

$$j \sim \rho v^3 \ln(\mathrm{Re}). \tag{9.36}$$

Multiplying this expression by the surface area of the pipe ($\sim Rl$), we obtain the dissipated power in turbulent flow:

$$\dot{E}' \sim jRl \sim \rho v^3 Rl \ln(\mathrm{Re}). \tag{9.37}$$

Comparing (9.35) and (9.37),

$$\frac{\dot{E}'}{\dot{E}} \sim \frac{v^3 R \ln(\mathrm{Re})}{\nu u^2} \sim \frac{\mathrm{Re}}{\ln^2(\mathrm{Re})} \gg 1. \tag{9.38}$$

Hence we conclude that *the rate of dissipation of the kinetic energy of the fluid into heat is much larger in turbulent flow than in laminar flow with the same velocity.*

Problem 1. Show that in the case of turbulent flow in a rough pipe the rate of energy dissipation is determined by the ratio of the size of an irregularity to the radius of the pipe, rather than by the absolute size of the irregularity.

Problem 2. Show that in the case of turbulent flow in a rough pipe if the size of a surface irregularity r_0 is large in comparison with the thickness y_0 of the viscous sublayer, then instead of (9.23) the logarithmic velocity profile has the form

$$v_x \sim v \ln(y/r_0).$$

Problem 3. Show that in the case of a body moving in a fluid with very large Reynolds number, turbulization of the boundary layer behind the body leads to a decrease in the drag force in comparison with the law $F \sim \rho u^2 R^2$. Here, R is the linear dimension of the body, u is the stream velocity, and ρ is the density of the fluid [compare (8.4)]. This phenomenon is called *crisis resistance.*

Problem 4. Explain why an increase in the compressibility of a fluid impedes turbulization of the boundary layer and therefore promotes stabilization of the flow.

Chapter 10

Heat transfer in liquids and gases

A flux of thermal energy is induced in a medium as a result of a temperature gradient. If the temperature gradient is small the heat flux density q is obviously proportional to the first power of the temperature gradient dT/dx, since this is the first term of a Taylor-series expansion of q in dT/dx. The linear relation is called **Fourier's law**:

$$q=-\lambda \frac{dT}{dx}. \tag{10.1}$$

The proportionality constant λ in (10.1) is the thermal conductivity [see (1.31)]; it was estimated for rarefied gases in Sec. 1.3.

In gases a temperature variation also leads to a variation in the density of the gas (this follows from the ideal-gas law for a rarefied gas, for example). The pressure is constant in such a process because equalization of the pressure is accomplished much more rapidly than transfer of thermal energy: equalization of pressure is accomplished by the macroscopic motion of the gas and the velocity of macroscopic motion is usually large in comparison with the typical drift velocity of the molecules in the process of heat conduction. The thermal energy of 1 kg of gas is $c_P T$, where c_P is the specific heat at constant pressure and is of order M^{-1}, where M is the mass of a gas molecule. Recall that the temperature is measured in energy units, i.e., the Boltzmann constant is equal to unity. The thermal energy of a unit volume of gas (or liquid) is equal to $\rho c_P T$, where ρ is the density of the material.

We write the heat balance equation, assuming that the heat flux is along x. The rate of change of thermal energy is obviously equal to $\partial(\rho c_P T)/\partial t$. On the other hand, this rate of change is determined by the difference $q_1 - q_2$ of heat fluxes going into (q_2) and out of (q_1) a unit volume of fluid through the opposing sides of this volume in the yz plane. Because the distance between these sides in the x direction is equal to unity, we obtain from (10.1)

$$q_2-q_1=\lambda \frac{\partial^2 T}{\partial x^2}.$$

Therefore the equation of conservation of thermal energy has the form

$$\rho c_P \frac{\partial T}{\partial t}=\lambda \frac{\partial^2 T}{\partial x^2}.$$

We introduce the thermal diffusivity $a = \lambda/(\rho c_P)$. Then the equation of conservation of thermal energy can be rewritten in the form

$$\frac{\partial T}{\partial t}=a\frac{\partial^2 T}{\partial x^2}. \tag{10.2}$$

In the case when heat is propagated along x, y, and z, we replace $\partial^2/\partial x^2$ in (10.2) by the Laplacian operator

$$\Delta=\frac{\partial^2}{\partial x^2}+\frac{\partial^2}{\partial y^2}+\frac{\partial^2}{\partial z^2}.$$

We then obtain the so-called **heat equation**

$$\frac{\partial T}{\partial t}=a\Delta T. \tag{10.3}$$

This equation holds for a fluid at rest. If the liquid or gas moves, then on the left-hand side of (10.3) the partial derivative $\partial T/\partial t$ must be replaced by the total derivative dT/dt. In this case the temperature $T(\mathbf{r},t)$ varies in time because the temperature varies at a given point of space $\mathbf{r}$ [this effect is taken into account in (10.3)], and also because when the fluid moves a given fluid element moves and its location in space is then occupied by a different fluid element with a different temperature. Therefore

$$\frac{dT}{dt}=\frac{\partial T}{\partial t}+(\mathbf{v}\nabla)T=a\Delta T. \tag{10.4}$$

We consider steady flow of the fluid, in which the temperature does not depend explicitly on time. In the case of a body moving with respect to a fluid, a steady problem results when we transform into a coordinate system in which the body is at rest and the fluid moves around it and gains or loses heat to the body. Then from (10.4) we find

$$(\mathbf{v}\nabla)T=a\Delta T. \tag{10.5}$$

The velocity of the fluid $\mathbf{v}$ in (10.5) is obtained from the solution of the Navier–Stokes equation (7.4) together with the incompressibility equation (6.1) (or the ideal-gas equation for a rarefied gas).

Besides the Reynolds number $\mathrm{Re} = vR/\nu$, motion of a fluid accompanied by heat transfer is characterized by a second dimensionless number $\mathrm{Pr} = \nu/a$, called the *Prandtl number*. Indeed, we see from (10.3) that the thermal diffusivity a has the dimensions m^2/s, which are the same dimensions as the kinematic viscosity ν.

Problem 1. Why does a pressure gradient in a gas not lead to heat transfer?

10.1. Propagation of heat in a medium

Suppose at the initial time a quantity of thermal energy Q is released at a certain point in the fluid. We estimate how the heated region expands in time and how the temperature inside this region falls off with time.

Let t be the time required for heat to propagate from the source to a region with linear dimension R. We obtain the following estimate from (10.3):

$$T/t \sim aT/R^2,$$

hence

$$R \sim (at)^{1/2}. \tag{10.6}$$

Therefore *the radius of the heated region expands as the square root of the time t.*

The relation (10.6) can be given another interpretation. If the material inside a volume with linear dimension R is heated nonuniformly, the temperature will equalize as a result of heat transfer after a time

$$\tau \sim R^2/a, \tag{10.7}$$

called the relaxation time.

We estimate how the temperature in the heated region falls off with time. Let Q be the quantity of heat released in the fluid at the initial time. From conservation of thermal energy inside a volume of order R^3, we obtain

$$Q \sim \rho c_P T R^3.$$

since the thermal energy per unit volume is equal to $\rho c_P T$. Hence, using (10.6), we find

$$T \sim \frac{Q}{\rho c_P} (at)^{-3/2}. \tag{10.8}$$

Therefore *in three-dimensional space the temperature falls off with time as* $t^{-3/2}$.

In the one-dimensional case, if we let Q be the quantity of heat released per unit area of a source filling the yz plane (perpendicular to the direction of propagation of the heat), then in analogy with (10.8) we have

$$T \sim \frac{Q}{\rho c_P} (at)^{-1/2}. \tag{10.9}$$

In this case the temperature falls off with time as $t^{-1/2}$. The two-dimensional case can be considered in the same way.

We note that the results obtained above are correct for solids, as well as liquids and gases. In the case of a solid or a liquid the difference between the specific heats at constant pressure and at constant volume is slight because of the small compressibility of the material.

We now consider a typical unsteady heat conduction problem. Let a plane source of heat have a temperature $T_0(t)$ which oscillates in time with frequency ω:

$$T_0(t) = T_0 \sin \omega t.$$

We estimate the characteristic distance x_0 over which the temperature wave generated by the heat source damps out.

We assume a solution in the form of a monochromatic plane wave:

$$T(x,t) = T_0 \exp(i\omega t - kx).$$

Substituting this solution into (10.2), we obtain the following result for the wave number k:

$$k=(i\omega/a)^{1/2}. \quad (10.10)$$

This solution satisfies the boundary condition of the problem at $x=0$ (note that obviously the temperature is real; however, because the heat equation (10.2) is linear, we can solve the equation using complex variables and take the real part of the solution at the end of the calculation).

We see from (10.10) that k is complex and its real and imaginary parts are equal. Therefore the wavelength $\lambda \sim (\mathrm{Re}\, k)^{-1}$ and the length $x_0 \sim (\mathrm{Im}\, k)^{-1}$ over which the wave damps out, are the same. This means that *propagation of a temperature wave is an aperiodic process in which the wavelength is equal to the penetration depth of the wave*:

$$x_0 \sim (a/\omega)^{1/2}. \quad (10.11)$$

Hence temperature waves do not actually exist as true waves. This statement is obviously independent of the geometry of the problem. In particular, (10.11) is also correct for a spherical source with an oscillating temperature.

Problem 1. The walls of a container with linear dimension R are maintained at a constant temperature T. An exothermic reaction occurs in the material inside the container, in which the thermal power released per unit volume is given by

$$j \sim \exp(-U/T).$$

Here, U is the activation energy. Show that two regimes can occur, depending on the value of j: either a steady temperature distribution is established inside the container or there is a "thermal explosion," in which there is a rapid unsteady heating of the material inside the container and a speeding up of the reaction process. Obtain the following value for the boundary value j_0 dividing these two regimes:

$$j_0 \sim \lambda T^2/(UR^2).$$

Here, λ is the thermal conductivity of the material.

10.2. Nonlinear heat conduction

In the preceding section we assumed that the thermal conductivity and thermal diffusivity were independent of temperature. As we saw in Sec. 1.3 [see (1.34)], for gases these coefficients depend on temperature in the form of a power law. In this section we consider heat conduction in a medium assuming that the thermal diffusivity is a power-law function of temperature with exponent n, i.e., $a = \alpha T^n$, where the coefficient α characterizes the medium.

We limit ourselves to the one-dimensional case. Let a plane source release a quantity of heat Q per unit area into the half space $x>0$ at time $t=0$. We first estimate how the heat propagation boundary moves away from the source and how the temperature varies with x near this boundary.

Let x_0 be the coordinate of the heat propagation boundary at time t. We consider the parameters on which x_0 depends, besides t. One of them, undoubtedly, is the constant α in the dependence of a on T. Since according to (10.2) the dimensions of a are m^2/s, it follows that $[\alpha] = m^2/[s\,(J)^n]$. Here we assume that the temperature is expressed in energy units (joules).

It would be incorrect to consider Q as a separate parameter on which x_0 depends and to construct a quantity with the dimensions of length from the parameters t, α, and Q. Since we are using the heat equation (10.2) for the temperature of the medium T, the initial conditions of the problem must also be expressed in terms of T. The distribution of thermal energy in space at the initial time $t=0$ has the form $Q\,\delta(x)$, where $\delta(x)$ is the one-dimensional δ function. Because Q has the dimensions J/m^2, and hence $Q\,\delta(x)$ has the dimensions of thermal energy density J/m^3, to determine the initial temperature distribution we must equate $Q\,\delta(x)$ to the thermal energy density $\rho c_P T$ at temperature T. Therefore the initial condition for the temperature T involves the combination $Q/(\rho c_P)$, and x_0 must depend on this same combination, which has the dimensions $[Q/(\rho c_P)] = J\,m$, since $[c_P] = kg^{-1}$.

Therefore to estimate x_0, we construct a quantity with the dimensions of length from t, α, and $Q/(\rho c_P)$. We first note that

$$\left[\alpha\left(\frac{Q}{\rho c_P}\right)^n\right] = m^{n+2}/\text{sec}.$$

Therefore

$$x_0^{n+2} \sim \alpha t \left(\frac{Q}{\rho c_P}\right)^n,$$

and so

$$x_0 \sim \left(\frac{Q}{\rho c_P}\right)^{n/(n+2)} (\alpha t)^{1/(n+2)}. \tag{10.12}$$

In particular, when $n=0$ (linear heat conduction) (10.12) reduces to (10.6), as expected. It follows from (10.12) that when $n>0$ the heated region grows in time more slowly than in the linear case $n=0$.

We next estimate the behavior of the temperature T near the boundary (10.12) of the heated region on the side closest to the source. It is first necessary to modify the heat equation (10.2) to apply to the case when a depends on T. Repeating the reasoning used in the introduction to Chap. 10, we find that the difference of the heat fluxes through opposing sides of a unit volume of the medium is

$$q_2 - q_1 \sim \frac{d}{dx}\left(\lambda \frac{dT}{dx}\right).$$

Therefore, assuming that ρ and c_P do not depend significantly on T, the equation describing the balance of thermal energy in the one-dimensional case has the form

$$\frac{\partial T}{\partial t} = \alpha \frac{\partial}{\partial x}\left(T^n \frac{\partial T}{\partial x}\right). \tag{10.13}$$

This equation generalizes (10.2) to the case of nonlinear heat conduction.

We estimate the terms of this equation for the region near the heat propagation boundary x_0. Let δx be the distance from x_0 in the direction of the heat source and assume that $\delta x \ll x_0$. Let v_0 be the velocity of the boundary x_0. It can be estimated as $v_0 = x_0/t$. Because $\delta x \ll x_0$, the velocity of heat propagation at the point $x_0 - \delta x$ is practically the same as the velocity of heat propagation at the boundary x_0. Therefore $\delta x = v_0\,\delta t$, where δt is the time required for the temperature to go from zero on the boundary to the value δT when the boundary moves to the point $x_0+\delta x$.

We find from (10.13)

$$\frac{\delta T}{\delta t} \sim \alpha \frac{(\delta T)^{n+1}}{(\delta x)^{-2}}. \tag{10.14}$$

Substituting $\delta t = \delta x/v_0$ and $v_0 \sim x_0/t$ in this relation, we obtain

$$\delta T \sim \left(\frac{x_0\,\delta x}{\alpha t}\right)^{1/n}. \tag{10.15}$$

Here, the boundary x_0 of the heat front is given by (10.12). Substituting (10.12) into (10.15), we determine the temperature δT as a function of the distance δx from the heat front boundary at time t:

$$\delta T \sim \left(\frac{Q}{\rho c_P}\right)^{1/n+2} (\delta x)^{1/n} (\alpha t)^{-(n+1)/n(n+2)}. \tag{10.16}$$

We see from (10.16) that when $n > 0$ the heat propagation boundary is well localized in space and the temperature on the boundary is strictly zero. If $n = 0$, then heat instantaneously propagates to infinity and the temperature is nonzero at all points of space for $t > 0$.

According to (10.16), the temperature increases with increasing δx as $(\delta x)^{1/n}$ at a fixed time t. Setting $\delta x \sim x_0$, we obtain the characteristic temperature T inside the heated region at time t. Using (10.12), we obtain from (10.16)

$$T \sim \left(\frac{Q}{\rho c_P}\right)^{2/(n+2)} (\alpha t)^{-1/(n+2)}. \tag{10.17}$$

In particular, when $n = 0$ we obtain the result (10.9), as expected.

We conclude from (10.17) that *the nonlinearity slows down the decrease of temperature with time when $n > 0$ and speeds it up when $n < 0$.* It follows from (10.17) that the results of this section are correct when $n > -2$.

Problem 1. A very large quantity of heat Q is released at a point at time $t = 0$ into a gas. Show that the radius r_0 of the front of heated gas increases in time as

$$r_0 \sim (\alpha Q^n t)^{1/(3n+2)}.$$

Here, n is the exponent in the dependence of the thermal diffusivity of the gas $a = \alpha T^n$ on temperature.

Problem 2. For the conditions of the preceding problem show that $n=5$ at very high temperatures in air.

10.3. Heat transfer of a body moving with respect to a fluid

In this section we consider heat transfer between a body and the surrounding fluid. We assume that the body has linear dimension R and the stream velocity of the fluid is equal to u. Let δT be the temperature difference between the body and the fluid, which also leads to heat transfer between the body and the fluid.

It follows from dimensional considerations that the temperature T at an arbitrary point in the fluid has the general form

$$T \sim \delta T\, f\left(\frac{r}{R}, \mathrm{Re}, \mathrm{Pr}\right).$$

Here, the function f depends on the shape of the body. The velocity of the fluid v has the simpler form

$$\mathbf{v} \sim \mathbf{u} \cdot F\left(\frac{r}{R}, \mathrm{Re}\right).$$

Here, the function F is also determined by the shape of the body. The Prandtl number Pr does not appear in the expression for $\mathbf{v}$ because the solution of the hydrodynamic equations for an incompressible fluid is independent of the solution of the heat equation. In gas dynamics the solutions are coupled if the density of the gas varies significantly in the process under consideration.

The *heat transfer coefficient* h is defined by the relation

$$h = q/\delta T, \tag{10.18}$$

where q is the heat flux density from the body to the fluid (or vice versa, depending on the sign of the temperature difference δT). The quantity q can be estimated from its definition (10.1):

$$q \sim \lambda\, \delta T/R. \tag{10.19}$$

Substituting (10.19) into (10.18), we obtain

$$h = \frac{\lambda}{R}\, \varphi\, (\mathrm{Re}, \mathrm{Pr}). \tag{10.20}$$

Here, φ is a dimensionless function of the dimensionless quantities Re and Pr.

The dimensionless number

$$\mathrm{Nu} = hR/\lambda$$

is called the *Nusselt number*. It follows from (10.20) that this number is a function only of the Reynolds number and the Prandtl number

$$\mathrm{Nu} = \varphi\, (\mathrm{Re}, \mathrm{Pr}).$$

Suppose that the Reynolds number is small, i.e., $\mathrm{Re} \ll 1$. Then, the dependence of φ on Re can be neglected in (10.20) because in this case the ratio of the left-hand side of (10.5) to the right-hand side is of order

$$(\mathbf{v}\nabla)T/(a\Delta T)\sim uR/a\ll 1.$$

Therefore the temperature T satisfies Laplace's equation

$$\Delta T=0, \tag{10.21}$$

as in the case of a fluid at rest. We note that this conclusion is correct when the Prandtl number Pr is sufficiently small, which was used when we neglected the left-hand side of (10.5):

$$uR/a\sim \text{Re}\cdot\text{Pr}\ll 1.$$

The solution of (10.21) is obviously independent of the kinematic viscosity ν. Therefore the heat transfer coefficient h does not depend on the Prandtl number Pr, which contains ν. It follows that when $\text{Re}\ll 1$ the function $\varphi(\text{Re},\text{Pr})$ in (10.20) is independent of both Re and Pr (even though Pr may not be small in comparison with unity). Then the function φ reduces to a numerical constant of order unity and we obtain from (10.20)

$$h\sim\lambda/R. \tag{10.22}$$

Hence we have estimated the heat transfer coefficient for small Reynolds numbers $\text{Re}\ll 1$ and arbitrary Prandtl number Pr.

We next estimate this coefficient for laminar flow past the body at large Reynolds number $\text{Re}\gg 1$ and $\text{Pr}\sim 1$. As we have mentioned above, a laminar boundary layer is formed in front of the body when $\text{Re}\gg 1$ and also behind the body when Re is less than the critical value $\text{Re}_{cr}\gg 1$. Because we have assumed that the Prandtl number $\text{Pr}=\nu/a$ is of order unity, it follows that the roles of the thermal conductivity and viscosity are comparable outside the boundary layer. Since we have neglected the viscosity, the thermal conductivity of the fluid over distances of order R can be neglected. The thermal conductivity leads to a heat transfer coefficient of order (10.22), whereas we will show below that the true heat transfer coefficient is much larger. Heat conduction actually occurs only inside a thin laminar boundary layer whose thickness is small in comparison with the linear dimension R of the body.

The heat flux density q through an area parallel to the surface of the body in the direction perpendicular to the surface is roughly

$$q=-\lambda\frac{dT}{dy}\sim\lambda\frac{\delta T}{\delta}.$$

Here, δ is the thickness of the laminar boundary layer and δT is the temperature difference between the body and the fluid. We obtain the following estimate for the heat transfer coefficient (10.18):

$$h\sim\lambda/\delta. \tag{10.23}$$

Substituting the thickness of the laminar boundary layer δ from (9.6) into (10.23), we find

$$h\sim\lambda\left(\frac{u}{R\nu}\right)^{1/2}\sim\frac{\lambda}{R}(\text{Re})^{1/2}\gg\frac{\lambda}{R}. \tag{10.24}$$

We see that the heat transfer coefficient at large velocities is much larger than at small velocities [compare (10.22)], which is obviously physically.

When $\mathrm{Re} \sim 1$ the expressions (10.22) and (10.24) are the same order of magnitude, as one would expect. In addition, it follows from (10.24) that when $\mathrm{Re} \gg 1$ the heat transfer coefficient is proportional to $R^{-1/2}$, whereas when $\mathrm{Re} \lesssim 1$ it is proportional to R^{-1}.

We now consider the heat transfer coefficient for laminar flow of a fluid past a body in the case $\mathrm{Re} \gg 1$ and $\mathrm{Pr} \gg 1$. As before, let δ be the thickness of the laminar boundary layer in the velocity [see (9.6)]. Let x be the direction parallel to the surface of the body and y be the direction perpendicular to the surface. Over the distance $y \sim \delta$ the velocity v_x of the fluid varies from zero to the stream velocity u of the fluid at infinity.

Furthermore, let $y \sim \delta'$ be the distance over which the temperature varies from the temperature of the body to the temperature of the fluid at infinity. Let δT be the difference between these temperatures. We will see below that $\delta' \ll \delta$.

We estimate the different terms in the steady heat equation (10.5) over the distance $y \sim \delta'$. The right-hand side of (10.5) is roughly

$$a \frac{\partial^2 T}{\partial y^2} \sim a \frac{\delta T}{\delta'^2}.$$

On the left-hand side of (10.5) the term

$$v_y \frac{\partial T}{\partial y} \sim v_y \frac{\delta T}{\delta'},$$

and the term

$$v_x \frac{\partial T}{\partial x} \sim v_x \frac{\delta T}{R}.$$

According to the incompressibility equation (6.1), when $y \sim \delta'$ we have

$$\frac{\partial v_x}{\partial x} \sim \frac{\partial v_y}{\partial y} \quad \text{or} \quad \frac{v_x}{R} \sim \frac{v_y}{\delta'}. \tag{10.25}$$

Therefore when $y \sim \delta'$ the two terms on the left-hand side of (10.5) are the same order of magnitude. Hence, qualitatively we can write (10.5) in the form

$$v_y \frac{\delta T}{\delta'} \sim a \frac{\delta T}{\delta'^2},$$

and therefore

$$v_y \sim a/\delta'. \tag{10.26}$$

We next estimate the velocities v_x and v_z at a distance $y \sim \delta'$ from the surface of the body. Because $v_x = 0$ at $y = 0$ and $v_x \sim u$ at $y \sim \delta$, if we expand v_x in a Taylor series in y, we will obtain for $y \sim \delta'$:

$$v_x \sim \frac{y}{\delta} u \sim \frac{\delta'}{\delta} u. \tag{10.27}$$

Such an expansion can be used in the case of a linear differential equation. In our problem we have the nonlinear term $(\mathbf{v}\cdot\nabla)\mathbf{v}$ in the Navier–Stokes equation (9.74). Therefore it must be shown that it can be neglected when $y \sim \delta'$. We have from (10.26)

$$\frac{(\mathbf{v}\cdot\nabla)\mathbf{v}}{\nu\Delta\mathbf{v}} \sim \frac{v_y}{\delta'}\frac{\delta'^2}{\nu} \sim \frac{1}{\mathrm{Pr}} \ll 1. \tag{10.28}$$

Substituting (10.27) into (10.25), we obtain the following estimate for the vertical component of the velocity v_y when $y \sim \delta'$

$$v_y \sim \delta'^2 u/(R\delta).$$

Substituting this estimate into (10.26), we obtain a relation between δ' and δ:

$$\delta'^3 u \sim a\,\delta R.$$

Hence using (9.6), we obtain

$$\delta' \sim \left(\frac{\delta a\, R}{u}\right)^{1/3} \sim \frac{\delta}{\mathrm{Pr}^{1/3}} \ll \delta. \tag{10.29}$$

Therefore we have shown that when the Prandtl number $\mathrm{Pr} \gg 1$, the temperature boundary layer is much thinner than the velocity boundary layer.

It also follows from (10.29) that when $\mathrm{Pr} \sim 1$ the quantities δ and δ' are of the same order of magnitude. This fact was used implicitly in finding the thickness of the laminar boundary layer and in obtaining the estimate (10.24).

Using (10.29), we can now estimate the heat transfer coefficient for the case $\mathrm{Re} \gg 1$, $\mathrm{Pr} \gg 1$, and where the flow is assumed to be laminar. Substituting δ' in place of δ in (10.23), since δ' determines the distance over which the temperature varies, we find

$$h \sim \frac{\lambda}{\delta'} \sim \frac{\lambda}{\delta}(\mathrm{Pr})^{1/3} \sim \frac{\lambda}{R}(\mathrm{Re})^{1/2}(\mathrm{Pr})^{1/3}. \tag{10.30}$$

We see that the rate of heat transfer increases with increasing Prandtl number and that the heat transfer coefficient increases as $\lambda^{2/3}$, where λ is the thermal conductivity of the fluid.

We make a general remark of a mathematical nature on the estimates of the derivatives used in the derivation of (10.30). If we have a function $f(x)$ whose variation is characterized by a linear dimension R, then when $x \sim R$ the derivative is roughly

$$\frac{df}{dx} \sim \frac{f(R)}{R}.$$

However, if we need to estimate $f(x)$ for $x \sim \delta \ll R$, then the estimate of the derivative will be different:

$$\frac{df}{dx} \sim \frac{f(\delta)}{\delta}.$$

In conclusion of this section we consider the process of heat transfer in the case of turbulent flow past a body when $\mathrm{Re} \gg 1$ and $\mathrm{Pr} \sim 1$.

As in the case of laminar flow, it is convenient to analyze the problem over a small local region of the boundary between the body and the fluid. Then the surface of the body in this region can be assumed to be approximately flat. As above, we let x be the direction of the main flow, parallel to the surface of the body, and y be the axis perpendicular to the surface. We consider a distance $y_0 \ll y \lesssim \delta$ from the surface. Let δ be the thickness of the turbulent boundary layer (9.26) and let y_0 be the thickness of the viscous sublayer (9.24).

The quantity $\rho c_P T$ is the thermal energy density in the region under consideration. The temperature T varies with height y. Hence the variation of the thermal energy density with thickness y of the turbulent boundary layer is given by $\rho c_P y\, dT/dy$. This variation produces a heat flux in the y direction (from the body to the fluid or vice versa). To estimate the heat flux recall that the vertical velocity component v_y does not depend on y and is of the order of the typical vortex velocity v of turbulent flow in the boundary layer [given by (9.22)].

Hence

$$q \sim v \rho c_P \frac{dT}{dy} y \tag{10.31}$$

is the heat flux density in the direction perpendicular to the surface of the body.

In the steady case q can be taken as given, i.e., it does not depend on y. Setting $q = \text{const}$ in (10.31), we obtain the dependence of the temperature difference δT across the turbulent boundary layer on the layer thickness δ by integrating (10.31) with respect to y from y_0 to δ:

$$\delta T \sim \frac{q}{\rho c_P v} \ln \frac{\delta}{y_0}. \tag{10.32}$$

Like the velocity component v_x (see Sec. 9.2), the temperature obeys a logarithmic law. The constant y_0 in (10.32) is determined from the condition (9.18). Therefore the entire temperature difference δT occurs over the thickness δ of the boundary layer, beginning with the minimum distance y_0.

The temperature variation in the viscous sublayer $y < y_0$ can be neglected, since in this sublayer the temperature difference is of the order of the factor in front of the logarithm in (10.32) and therefore we can neglect it in comparison to (10.32) since $\delta \gg y_0$.

Substituting (10.32) into the general expression (10.18) for the heat transfer coefficient, and using (9.22), (9.24), and (9.26), we find

$$h \sim \frac{\rho c_P v}{\ln(\mathrm{Re})} \sim \frac{\rho c_P u}{\ln^2(\mathrm{Re})}. \tag{10.33}$$

Here, the Reynolds number $\mathrm{Re} = uR/\nu \gg 1$.

We compare (10.33) (let h in this case be denoted as h_t) with the expression (10.24) for the heat transfer coefficient in the case of laminar flow with $\mathrm{Re} \gg 1$, $\mathrm{Pr} \sim 1$ (let h in this case be denoted as h_l). Because the thermal conductivity $\lambda = \rho c_P a$, we find, dividing (10.33) by (10.24),

$$\frac{h_t}{h_l} \sim \frac{uR}{a(\mathrm{Re})^{1/2} \ln^2(\mathrm{Re})} \sim \frac{\mathrm{Pr}\cdot(\mathrm{Re})^{1/2}}{\ln^2(\mathrm{Re})} \gg 1. \qquad (10.34)$$

Therefore we conclude that for the same values of the parameters *the rate of heat transfer through a turbulent boundary layer is much larger than through a laminar boundary layer.* This conclusion remains valid for an arbitrary value of the Prandtl number.

Problem 1. Show that the heat transfer coefficient for turbulent flow with $\mathrm{Re} \gg 1$ and $\mathrm{Pr} \gg 1$ is proportional to $\mathrm{Pr}^{3/4}$.

Problem 2. Show that the amplitude of the temperature oscillations in vortices of turbulent flow with linear dimensions $l_0 \ll l < R$ [l_0 is the smallest scale of length of the turbulent flow; see (8.6)] is proportional to $l^{1/3}$, as in the case of the velocity oscillations [see (8.5)]. Assume that the Prandtl number is of order unity.

Problem 3. Consider Poiseuille flow in a pipe of radius R and length l under a pressure difference δP. The temperature of the surface of the pipe is held constant. Show that the temperature difference between points near the axis of the pipe and at its surface is of order

$$\delta T \sim (\delta P\, R^2)^2/(\lambda l^2 \eta).$$

10.4. Heating of a body in a moving fluid

We consider the heating of a body moving with a constant velocity in a fluid. Let R be the linear dimension of the body and u be its velocity. First, we assume that the velocity is not too large and so the Reynolds number $\mathrm{Re} = uR/\nu \ll 1$.

Up to now, in writing the heat equation in the form (10.4), we have neglected the transformation of kinetic energy of the fluid into heat. We obtain the condition under which this approximation is valid. According to (7.40) and (7.42), the dissipated power per unit volume of fluid is of order

$$j \sim \eta \frac{u^2}{\delta^2}. \qquad (10.35)$$

Here, δ is the thickness of the layer surrounding the surface of the body in which the velocity of the fluid increases from zero to the stream velocity u. When $\mathrm{Re} \ll 1$ a boundary layer does not exist and $\delta \sim R$.

To neglect this source of energy dissipation, (10.35) must be small in comparison with the rate of change of thermal energy per unit volume due to the heat flux. According to (10.4), this rate of change is of order

$$j' \sim \lambda \frac{\delta T}{R^2}. \qquad (10.36)$$

Dividing (10.35) by (10.36), we obtain

$$\frac{j}{j'} \sim \mathrm{Pr} \frac{u^2}{c_P\, \delta T}. \qquad (10.37)$$

This ratio is small in comparison with unity if the kinetic energy per unit mass of fluid ($\sim u^2$) is small in comparison with the thermal energy of the fluid $c_P \delta T$. This condition will be true for small Reynolds numbers if δT is not too small. For example, if $\delta T \sim T$, then

$$c_P \delta T \sim c_P T \sim \frac{1}{M} M v_t^2 \sim v_t^2, \tag{10.38}$$

and we find that the ratio (10.37) is of order $(u/v_t)^2$, where v_t is the average thermal velocity of the molecules. Therefore the dissipation of the kinetic energy can be neglected if the velocity u of the fluid is small in comparison with the thermal velocities of the molecules of the fluid. This condition does not require small Reynolds number, but can also be satisfied for $\mathrm{Re} \gg 1$, in particular, for fully developed turbulence.

However, if the value of δT is unknown (unlike the problems of the preceding section of this chapter), then the term (10.35) must be retained: it determines the temperature difference δT between the body and the fluid. Since $\delta T \ll T$, the dissipated powers (10.35) and (10.36) can be of the same order of magnitude. Equating their ratio (10.37) to unity, we obtain

$$\delta T \sim \mathrm{Pr}\, M u^2. \tag{10.39}$$

Here, we have again used the fact that the specific heat of a fluid $c_P \sim M^{-1}$, where M is the mass of a molecule of the fluid. Finally, the above results are valid for both liquids and gases.

Hence the temperature difference between the body and the fluid is proportional to the square of the velocity of the fluid with respect to the body. The proportionality constant in (10.39) depends on the shape of the body. For example, it is equal to $\frac{5}{8}$ for a sphere. Note that δT in (10.39) does not depend on the size of the body R.

Next we estimate the rate of heating of the body for large Reynolds number $\mathrm{Re} = uR/\nu \gg 1$. We first consider laminar flow, which takes place everywhere when $1 \ll \mathrm{Re} \ll \mathrm{Re}_{\mathrm{cr}}$ or in front of the body only when $\mathrm{Re} > \mathrm{Re}_{\mathrm{cr}}$. In (10.35) for the dissipated power per unit volume it is necessary to substitute the thickness δ of the velocity boundary layer (9.6). Here, we assume that the Prandtl number Pr is arbitrary. If $\mathrm{Pr} \gg 1$ then the thickness δ' of the temperature boundary layer is much smaller than that of the velocity boundary layer [see (10.29)]; then it is incorrect to equate (10.35) to the change in the heat flux. Indeed, over the thickness δ' the dissipated power is small in comparison with the total dissipated power. The correct approach is to equate the powers per unit surface area, and not per unit volume.

In the case of small Reynolds number this approach again leads to the estimate (10.39), as it must. However, for large Reynolds numbers the situation is as follows. The power arriving on a unit area of the surface of the body is given by (10.35) as

$$j\delta \sim \eta u^2/\delta.$$

It is transformed into a heat flux q per unit surface area of the body, i.e., a heat flux density

$$q=\lambda\frac{dT}{dy}\sim\lambda\frac{\delta T}{\delta'}. \qquad (10.40)$$

The quantity δ' is given by (10.29). Hence the balance of energy per unit surface area of the body has the form

$$\frac{\eta u^2}{\delta}\sim\frac{\lambda\,\delta T}{\delta'}. \qquad (10.41)$$

Using (10.29), we obtain an estimate of the induced temperature difference δT for $\mathrm{Re}\gg 1$:

$$\delta T\sim(\mathrm{Pr})^{2/3}\,Mu^2. \qquad (10.42)$$

In particular, when the Prandtl number $\mathrm{Pr}\sim 1$ the expression (10.42) has the same form as (10.39) for small Reynolds numbers $\mathrm{Re}\ll 1$. Comparing (10.39) and (10.42), we conclude that only the functional dependence of the induced temperature difference on the Prandtl number Pr changes as the velocity of the fluid increases and the Reynolds number goes from small to large values (but the flow remains laminar).

Finally, we consider the heating of the body in the case of turbulent flow with $\mathrm{Re}\gg 1$ and $\mathrm{Pr}\sim 1$. The dissipated power per unit surface area of the body is, according to (9.17) and (9.22),

$$\dot{E}\sim\rho v^3\ln(\mathrm{Re})\sim\frac{\rho u^3}{\ln^2(\mathrm{Re})}. \qquad (10.43)$$

We set (10.43) equal to the heat flux density (10.40), where δ' is the thickness of the temperature boundary layer in turbulent flow. Because the Prandtl number Pr is of order unity, the thickness δ' is comparable to the thickness δ of the velocity boundary layer (9.26) for turbulent flow. We then obtain for the induced temperature difference

$$\delta T\sim\frac{\rho u^3\delta}{\lambda\ln^2(\mathrm{Re})}\sim\frac{\mathrm{Re}}{\ln^3(\mathrm{Re})}Mu^2. \qquad (10.44)$$

We compare (10.44) with (10.42), setting $\mathrm{Pr}\sim 1$ in (10.42). We see that the heating of the body in turbulent flow is a factor of Re more intense than in laminar flow, neglecting the weak logarithmic dependence on the Reynolds number in (10.44). For example, we conclude from the above results that when a body moves in a fluid with a high velocity the back of the body is heated more strongly than the front of the body.

Problem 1. Generalize the result (10.44) for the induced temperature difference to the case of large Prandtl numbers for turbulent flow.

Chapter 11

Convection and diffusion

If the lower regions of a fluid are heated more strongly than the upper regions, then the material expands from the heating and the density in the lower regions becomes smaller than that in the upper regions. According to Archimedes's principle the lighter fluid rises and we have a macroscopic flow of hotter fluid upward and colder fluid downward. This phenomenon is called convection. Because of the relative macroscopic motion, the temperature of the fluid tends to become equalized. In general this process can transfer heat upward at a faster rate than the process of conduction. We note that in convection the material can be assumed to be incompressible, and so all of the results given below hold for both liquids and gases.

Convection occurs at constant pressure, since pressure is equalized much more rapidly than the convection process: equalization of pressure is due to the elastic forces in the material, whereas convection is due to buoyancy forces, which are small because of the small temperature difference, and therefore density difference, between the cold and hot regions of the fluid.

We generalize the Navier–Stokes equation to include buoyancy forces. We assume steady flow. The buoyancy force acting on a unit volume of material is equal to $\mathbf{F} = \rho'\mathbf{g}$, where ρ' is the density difference between the hot and cold regions of the fluid and $\mathbf{g}$ is the acceleration of gravity. The small density difference ρ' can be expressed through the small temperature difference T' between the hot and cold regions of the fluid:

$$\rho' = \left(\frac{\partial \rho}{\partial T}\right)_p T'.$$

The quantity

$$\beta = -\frac{1}{\rho}\left(\frac{\partial \rho}{\partial T}\right)_p \tag{11.1}$$

is called *the volume coefficient of expansion* of the fluid. Therefore we obtain for the buoyancy force

$$\mathbf{F} = -\beta \mathbf{g} \rho T'.$$

This force must be added to the right-hand side of the steady Navier–Stokes equation (7.4):

$$(\mathbf{v}\nabla)\mathbf{v} = -\frac{1}{\rho}\nabla P' + \nu\Delta\mathbf{v} - \beta\mathbf{g}T'. \tag{11.2}$$

Here, the pressure P' does not include the static pressure $P = \rho gz$, which is balanced by gravity $(\nabla P = \rho\mathbf{g})$.

The incompressibility equation of the fluid (6.1) must be satisfied, in addition to (11.2). For the conditions considered here (6.1) also holds for gases. Finally, the system of equations is closed by the heat equation (10.5). Convection in a gravitational field is called free convection.

Problem 1. Show that free convection arises if the temperature decreases in the upward direction and the temperature gradient is larger in absolute value than the quantity $\beta gT/c_P$. Here, c_P is the specific heat of the fluid at constant pressure. Also, show that in the case of an ideal gas this quantity is of order Mg, where M is the mass of a gas molecule.

11.1. Free convection of a hot liquid

We consider the simplest convective process in a liquid, when the surface of the liquid has a lower temperature than the bottom of the container. Unlike the problems considered earlier, here the characteristic velocity $\mathbf{v}$ of convective flow is not given: it must be found from (11.2), (6.1), and (10.5). The Reynolds number $\mathrm{Re} = vH/\nu$ then determines the nature of free convective motion: laminar $(\mathrm{Re} < \mathrm{Re}_{cr} \gg 1)$ or turbulent $(\mathrm{Re} > \mathrm{Re}_{cr})$. Here H is the depth of the liquid and ν is its kinematic viscosity.

We construct a dimensionless quantity analogous to the Reynolds number from the parameters characterizing the process of free convection. In addition to H and ν we also have the thermal diffusivity a of the liquid and the temperature difference δT between the bottom and the surface. We assume that this temperature difference is maintained by external sources. We see from (11.2) that the quantity δT does not appear directly in the problem, but only in the combination $\beta g\delta T$, which has the dimensions of $\mathrm{m/s^2}$. The following dimensionless quantity can be constructed from these parameters:

$$\mathcal{R} = \frac{H^3\beta g\,\delta T}{\nu a}, \tag{11.3}$$

which is called *the Rayleigh number.*

A second dimensionless parameter is obviously the Prandtl number $\mathrm{Pr} = \nu/a$ introduced in the preceding chapter. The number Pr depends only on the intrinsic properties of the liquid. Note that the density of the liquid ρ does not appear separately in (11.3).

From similar dimensional considerations the velocity of free convective motion can be written in the form

$$v = \frac{\nu}{H} f(\mathcal{R}). \tag{11.4}$$

Here, f is a function of the Rayleigh number, which cannot be obtained from general dimensional considerations because the Rayleigh number itself is dimensionless.

Next, we estimate the heat transfer coefficient h from the bottom of the liquid to its surface in free convection. The heat flux density in the vertical direction is approximately

$$\mathbf{q} = -\lambda \nabla T' \sim \lambda\, \delta T/\delta.$$

Here, δ is the distance in the vertical direction over which the temperature difference δT occurs in the liquid. It is not necessarily equal to the depth of the liquid because the entire temperature variation could occur in a thin layer near the bottom, rather than uniformly over the entire depth of the liquid.

We obtain from the heat equation [see (10.5)]

$$v\, \delta T/\delta \sim a\, \delta T/\delta^2.$$

Therefore

$$\delta \sim a/v. \tag{11.5}$$

It is evident from (11.5) that when the convective velocity is large the value of δ can indeed be small in comparison with the depth H.

Substituting (11.4) into (11.5) and assuming that the Prandtl number $\mathrm{Pr} \sim 1$ (as we will assume in all of the problems considered in this chapter for simplicity), we obtain

$$\delta \sim \frac{H}{f(\mathscr{R})}. \tag{11.6}$$

Using (11.6), we obtain for the heat flux density

$$q \sim \frac{\lambda\, \delta T}{H} f(\mathscr{R}). \tag{11.7}$$

The heat transfer coefficient h for convection is defined in the same way as for heat conduction [compare (10.18)]:

$$h = q/\delta T.$$

Then we obtain from (11.7) the following estimate for h:

$$h \sim \frac{\lambda}{H} f(\mathscr{R}). \tag{11.8}$$

If the Rayleigh number $\mathscr{R} \lesssim 1$, then we obviously have $f(\mathscr{R}) \sim 1$. Therefore

$$h \sim \lambda/H,$$

and so the heat transfer coefficient is the same order of magnitude as the expression (10.22) in the case of heat conduction. Therefore *when $\mathscr{R} \lesssim 1$ convection is not important and heat is transferred by conduction. However, if $\mathscr{R} \gg 1$ then convection dominates over conduction.*

In the limiting case of large Rayleigh number $\mathscr{R} \gg 1$ the convective motion is turbulent. As we saw in Chap. 8, the viscous term in (11.2) can be neglected in this

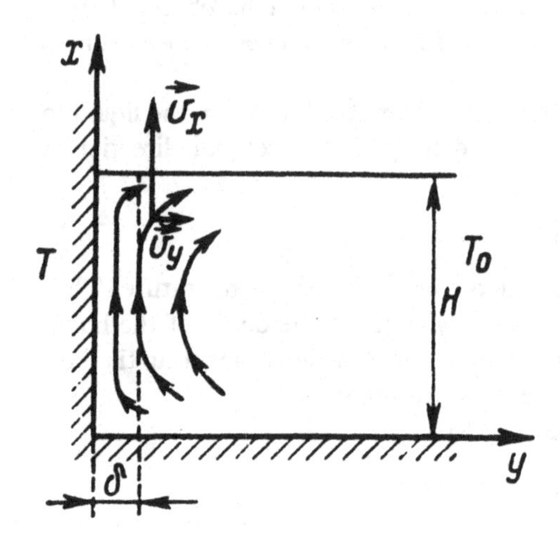

FIG. 17. Streamlines for free convection induced by a heated vertical wall. The dashed vertical line marks the end of the boundary layer, δ is the thickness of this layer, H is the depth of the liquid, T is the temperature of the wall, and T_0 is the temperature of the liquid.

case. This implies that the velocity (11.4) cannot depend on ν, which according to (11.3) will be true if $f(\mathscr{R}) \sim \mathscr{R}$. Then from (11.8) we obtain the heat transfer coefficient in the form

$$h \sim \beta g \, \delta T \rho c_p H^2/\nu.$$

In the limiting case $\mathscr{R} \gg 1$ it follows from (11.4) that the Reynolds number $\mathrm{Re} = Hv/\nu \sim \mathscr{R}$ and so in convection problems the Rayleigh number plays the role of the Reynolds number in determining the nature of the flow. In convection problems the flow transforms from laminar to turbulent for a certain critical value of the Rayleigh number $\mathscr{R}_{\mathrm{cr}} \gg 1$, corresponding to the critical Reynolds number $\mathrm{Re}_{\mathrm{cr}} \gg 1$.

We next consider a second typical problem of free convection. We estimate the heat transfer coefficient for the case of a vertical wall of height H heated to a temperature T and forming a boundary with a liquid whose temperature far from the wall is equal to T_0. We assume that the Rayleigh number is large, which implies that free convective motion is strong and we can neglect ordinary heat conduction.

We direct the x axis upward along the wall and the y axis in the horizontal direction perpendicular to the wall (Fig. 17). From the incompressibility equation (6.1) it follows that

$$v_x/x \sim v_y/y.$$

We will see below that the typical horizontal distance y over which heat is transported is small in comparison with the height H characterizing the scale of length along x. Therefore we will see that heat transfer occurs in a thin boundary layer near the wall and so $v_y \ll v_x$.

The heat equation (10.5) is approximately

$$v_x \frac{\partial T}{\partial x} \sim a \frac{\partial^2 T}{\partial y^2}. \tag{11.9}$$

Here, we have neglected the term $v_y \, \partial T/\partial y$ on the left-hand side of (10.5) because, according to the incompressibility equation (6.1), it is the same order as the term $v_x \, \partial T/\partial x$, which is included on the left-hand side of (11.9). We have also used the fact that on the right-hand side of (10.5) the derivative with respect to y is large in comparison with the derivative with respect to x. Assuming that the Prandtl number $\mathrm{Pr} = \nu/a$ is of order unity, we find the vertical velocity component v_x from (11.9)

$$v_x \sim H\nu/y^2. \tag{11.10}$$

We next consider the terms in the x component of the vector equation (11.2):

$$v_x \frac{\partial v_x}{\partial x} \sim \nu \frac{\partial^2 v_x}{\partial y^2} \sim \beta g(T - T_0). \tag{11.11}$$

We have neglected the term $\partial P'/(\rho\, \partial x)$ in (11.2) because the differentiation with respect to x gives the small factor H^{-1} [the pressure P' is determined from the y component of (11.2), where the term $\partial P'/(\rho\, \partial y)$ is of the same order of magnitude as the remaining terms]. The far left term of (11.11) is roughly v_x^2/H, while the middle term is roughly $\nu v_x/y^2$. It follows from (11.10) for v_x that the far left and middle terms of (11.11) are the same order of magnitude. We can then rewrite (11.11) in the form

$$\nu \frac{v_x}{y^2} \sim \beta g(T - T_0). \tag{11.12}$$

Substituting (11.10) into (11.12), we obtain the relation

$$H \frac{\nu^2}{y^4} \sim \beta g(T - T_0). \tag{11.13}$$

This result gives the thickness of the boundary layer $y = \delta$ over which heat transfer occurs. Using the definition of the Rayleigh number (11.3), we find

$$\delta \sim \frac{H}{(\mathscr{R})^{1/4}} \ll H. \tag{11.14}$$

The inequality in (11.14) follows from the condition $\mathscr{R} \gg 1$. Hence we have shown that *the temperature varies in a thin boundary layer near the surface of the wall.*

The mechanism of convection is as follows. The lower regions of hot liquid rise according to Archimedes's principle and are deflected slightly sideways into the colder regions of the liquid (see Fig. 17). The cold liquid in the lower regions outside the boundary layer is drawn into the space vacated by the rising liquid, is heated up by the wall, and the process is repeated, leading to a steady-state convective flow of the liquid.

Substituting (11.4) into (11.10), we find the typical velocity v_x of the vertical convective motion of the liquid:

$$v_x \sim \frac{\nu}{H} (\mathscr{R})^{1/2}. \tag{11.15}$$

The typical velocity in the horizontal direction v_y (the velocity with which the liquid is deflected sideways in rising) is of order

$$v_y \sim \frac{\nu}{H} (\mathscr{R})^{1/4}. \tag{11.16}$$

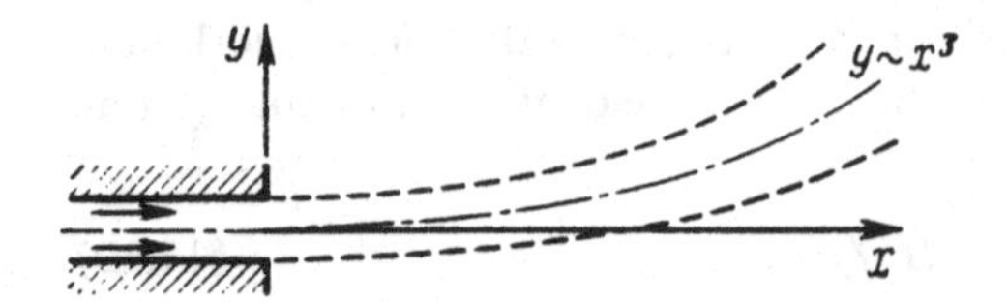

FIG. 18. Hot turbulent jet of gas ejected from an aperture and rising under convection.

The results (11.15) and (11.16) are consistent with the general formula (11.4) for the velocity of a fluid in free convection.

The quantity ν/H represents the typical velocity in this problem due to ordinary conduction. This is obvious from simple dimensional considerations. *Hence when* $\mathcal{R} \gg 1$ *the velocity of convective motion (11.15)–(11.16) is much larger than the rate of heat transfer by means of conduction.* This confirms the conclusion stated above that convection dominates over conduction for large Rayleigh number.

We estimate the typical Reynolds number in this problem. We find from (11.15)

$$\mathrm{Re} = \frac{v_x H}{\nu} \sim \mathcal{R}^{1/2} \gg 1. \tag{11.17}$$

As before, we see that in this problem the Rayleigh number plays the role of the Reynolds number.

We next estimate the heat transfer coefficient h. The heat flux density q from a unit area of the surface of the wall is given by (10.1). We have the estimate

$$q \sim \lambda \frac{T - T_0}{\delta}, \tag{11.18}$$

where λ is the thermal conductivity of the fluid. Then, using (10.18), we obtain an estimate for the heat transfer coefficient

$$h \sim \lambda/\delta.$$

Substituting (11.14) for δ in the above relation, we find

$$h \sim \frac{\lambda}{H} (\mathcal{R})^{1/4}. \tag{11.19}$$

If $\mathcal{R} \lesssim 1$ then there is no convection and heat is transferred by means of conduction. According to (10.22), we have $h \sim \lambda/H$ in this case. Comparing this expression with (11.19), we see that when $\mathcal{R} \gg 1$ convective heat transfer is much stronger than transfer by means of conduction. It follows from (11.8) and (11.19) that in this particular problem the function f has the form $f(s) \sim s^{1/4}$.

Problem 1. For the convective flow shown in Fig. 17, obtain the following estimate for the excess pressure P'

$$P' \sim \rho \left(\frac{\nu}{H}\right)^2 (\mathcal{R})^{1/2}. \tag{11.20}$$

Problem 2. A hot turbulent jet of gas spurts out horizontally from an aperture and is deflected slightly upward under the buoyancy force (Fig. 18). Show that the jet

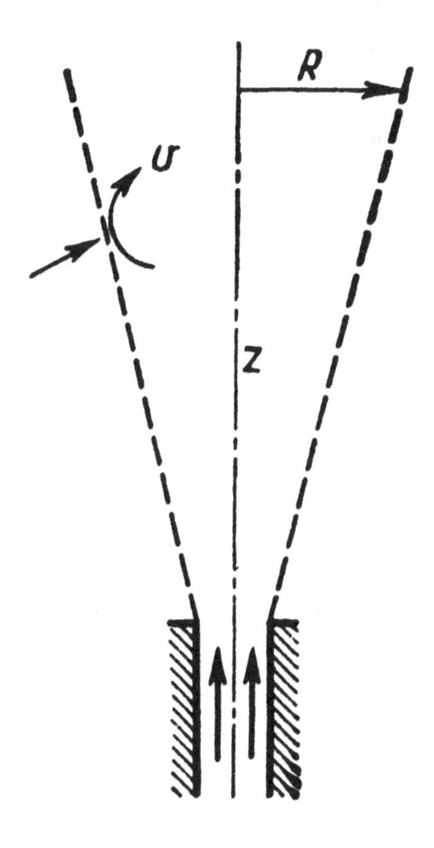

FIG. 19. Turbulent convection of a gas from a pipe. The dashed lines show the boundaries of the turbulent column of gas; R is the radius of the column at a height z above the pipe.

has the shape of a cubic parabola.

11.2. Rising column of heated gas

We consider a hot gas flowing out of a pipe. We asusme that the Rayleigh number is sufficiently large, so that the gas column rises due to convection

We first consider the case when the Rayleigh number is so large that the flow is turbulent. Let R be the radius of the column at a height z above the pipe opening (Fig. 19). In the case of fully developed turbulence we have $R \sim z$ [see (8.9)], since R cannot depend on the kinematic viscosity ν, and therefore R and z cannot be connected by any other relation (for example, containing the Rayleigh number).

In the case of turbulent motion all of the velocity components of the gas are the same order of magnitude, which we denote by $v = v(z)$. The total heat flux carried by the motion of the heated gas must be constant:

$$Q \sim \rho c_p T' v R^2 \sim \rho c_p T' v z^2 = \text{const.} \tag{11.21}$$

Here, $T'(z)$ is the temperature difference between the column and the surrounding cold air, and c_p is the specific heat of the gas at constant pressure. Recall that convection is a constant-pressure process. We take the z component of the vector equation (11.2). The term involving the viscosity $\nu \Delta \mathbf{v} \sim \nu v/z^2$ is small in comparison with the term $(\mathbf{v}\cdot\nabla)\mathbf{v} \sim v^2/z$, since their ratio is of order vz/ν, i.e., the

Reynolds number, which is large in comparison with unity in turbulent flow. Therefore the viscosity can be neglected for turbulent flow (see the detailed discussion in Chap. 8).

Furthermore, because in turbulent flow the pressure P' reduces to the kinematic pressure ρv^2 [see (8.4)], it follows that the term $\nabla P'/\rho$ in (11.2) is of order $P'/\rho z \approx v^2/z$, i.e., it is of the same order of magnitude as the left-hand side $(\mathbf{v}\cdot\nabla)\mathbf{v}$ of (11.2).

Hence the z component of (11.2) leads to the relation

$$v^2/z \sim \beta g T', \tag{11.22}$$

and therefore we obtain

$$v^2(z) \sim \beta g z T'(z). \tag{11.23}$$

Eliminating the velocity v from (11.21) and (11.23), we find the temperature difference T':

$$T'(z) \sim \frac{Q^{2/3}}{(\beta g \rho^2 c_p^2)^{1/3}} \frac{1}{z^{5/3}}. \tag{11.24}$$

Hence the temperature difference decreases with height z as $z^{-5/3}$. Substituting (11.24) into (11.23), we find the velocity v of the turbulent flow as a function of the height z:

$$v(z) \sim \left(\frac{\beta g Q}{\rho c_p}\right)^{1/3} \frac{1}{z^{1/3}}. \tag{11.25}$$

We see that the velocity decreases as $z^{-1/3}$.

The Rayleigh number (11.3) is estimated as

$$\mathscr{R} \sim \frac{z^3 \beta g T'}{\nu a} \sim \left(\frac{zv}{\nu}\right)^2 = \mathrm{Re}^2 \sim \left(\frac{\beta g Q}{\rho c_p}\right)^{2/3} \frac{z^{4/3}}{\nu^2} \gg 1. \tag{11.26}$$

It increases with height z as $z^{4/3}$ and therefore the flow becomes ever more turbulent with increasing height. The condition $\mathscr{R} \gg 1$ corresponds to $\mathrm{Re} \gg 1$, which in turn corresponds to heights

$$z \gg \left(\frac{\nu^3 \rho c_p}{\beta g Q}\right)^{1/2}, \tag{11.27}$$

and the solution obtained above is correct for heights satisfying this condition.

More exactly, turbulence sets in when $\mathrm{Re} > \mathrm{Re}_{cr} \gg 1$. Therefore there exists a region of Rayleigh numbers $1 \ll \mathscr{R} < \mathscr{R}_{cr}$ for which the convective motion of the gas is laminar. We turn now to laminar convection.

Again let R be the radius of the rising column of gas at a height z from the pipe. Here, we cannot assume $R \sim z$, as in the case of turbulent motion. We will see below that $R \ll z$. The condition of constant heat flux takes a somewhat different form from (11.21):

$$Q \sim \rho c_p T' v_z R^2 = \text{const.} \tag{11.28}$$

Here, v_z is the vertical component of the velocity of the gas.

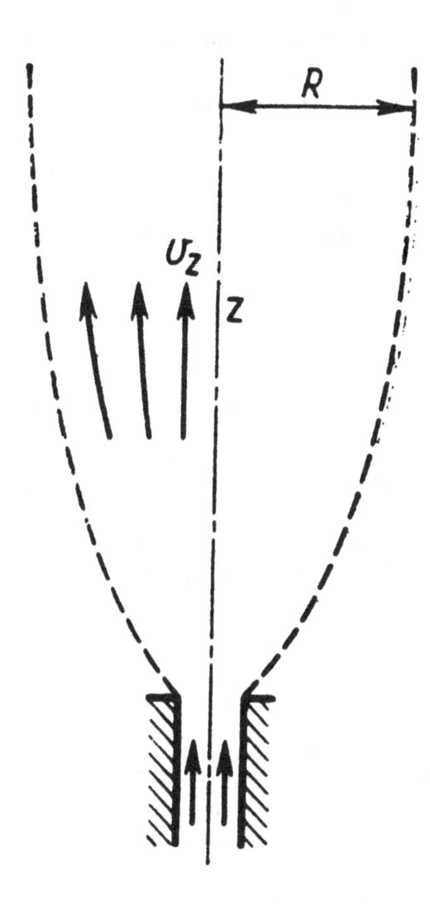

FIG. 20. Laminar convection of gas from a pipe. The dashed lines represent the boundary of laminar flow. The streamlines of the gas are also shown.

Taking the z component of the vector equation (11.2), we obtain

$$\frac{v_z^2}{z} \sim \nu \frac{v_z}{R^2} \sim \beta g T'. \tag{11.29}$$

Eliminating T' from (11.28) and (11.29), we find that the velocity component v_z does not depend on z and is roughly

$$v_z \sim \left(\frac{\beta g Q}{\rho c_p \nu}\right)^{1/2} = \text{const}, \tag{11.30}$$

which is the same order of magnitude as the velocity at which the gas is ejected from the pipe.

Equating the first two terms of (11.29), we obtain a relation between R and z (Fig. 20):

$$R \sim \left(\frac{\nu z}{v_z}\right)^{1/2}. \tag{11.31}$$

Here, the constant quantity v_z is found from (11.30). The condition $R \ll z$ is satisfied for sufficiently large z, in particular when $z \gg \nu / v_z$. Substituting (11.30) for

v_z into this inequality, it is easy to verify that it takes the form (11.27), which corresponds to Rayleigh numbers $\mathcal{R} \gg 1$.

Using (11.28) and (11.30), we obtain an estimate for the temperature T':

$$T'(z) \sim \frac{Q}{\rho c_p v_z R^2} \sim \frac{Q}{\rho \nu c_p z}. \tag{11.32}$$

The horizontal velocity components v_x and v_y can be estimated by using the incompressibility equation (6.1):

$$\frac{v_x}{R} \sim \frac{v_y}{R} \sim \frac{v_z}{z}. \tag{11.33}$$

Using (11.31) and (11.33), we find

$$v_x, v_y \sim \left(\frac{\nu v_z}{z}\right)^{1/2} \ll v_z. \tag{11.34}$$

The Rayleigh number is defined with respect to the distance R, rather than z, in analogy with the Reynolds number in the laminar wake problem. Hence we obtain (compare Sec. 7.3)

$$\mathcal{R} = \frac{R^3 \beta g T'}{\nu^2} \sim \left(\frac{z v_z}{\nu}\right)^{1/2} \gg 1. \tag{11.35}$$

Here, we have also used (11.31) and (11.32).

We see from (11.35) that the Rayleigh number $\mathcal{R}$ increases as $z^{1/2}$ until it reaches the critical value $\mathcal{R}_{cr}$, at which the flow goes from laminar to turbulent. According to (11.26), the Rayleigh number increases more rapidly with further increase in z (as $z^{4/3}$).

In the laminar regime the radius of the column increases slowly with z [as $z^{1/2}$; see (11.31)]. After the flow becomes turbulent the radius R increases much more rapidly with z (proportional to the first power of z).

The velocity v_z of the convective rising of the column is at first constant [see (11.30)] and then begins to fall off as $z^{-1/3}$ when the motion becomes turbulent [see (11.25)].

Finally, all of the qualitative results of this section are applicable not only to the flow of a gas from a pipe, but also to the convective rise of a gas jet from a hot body.

Problem 1. A plane convective jet of gas rises from a hot wire lying horizontally. Show that the velocity of the rising turbulent jet is independent of the height z above the wire, while the temperature difference T' between the jet and the surrounding air falls off with height as z^{-1}.

11.3. Diffusion of particles in a fluid

We have already considered the diffusion of gases in Sec. 1.2 and have estimated the diffusion coefficient. In this section we consider the general macroscopic features of the diffusion process, and we also estimate the diffusion coefficient for macroscopic particles diffusing in a fluid. Finally, we consider the relation between this process and the diffusion of molecules of a heavy gas in a light gas, as considered in Sec. 1.2.

Let c be the ratio of the mass of the diffusing material to the mass of the fluid in a given volume element. This dimensionless quantity is called the *concentration* (or mass concentration). We will assume that it is small, i.e., $c \ll 1$. Let j be the mass flux density of the diffusing material. The quantity j is determined not by the concentration c itself, but by its gradient dc/dx in the direction of the flux. For small gradients, j can be expanded in a Taylor series in dc/dx. The first term of this series gives

$$j = -\rho D \frac{dc}{dx}. \tag{11.36}$$

Here, ρ is the density of the fluid in which the material diffuses. The quantity D is called the *diffusion coefficient* [see also (1.6)]. The definitions (1.6) and (11.36) are the same, since $j = Mi$, where i is the flux density of particles of the diffusing material, and M is the mass of a particle.

As already noted in Sec. 1.2, it follows from (11.36) that the dimensions of the diffusion coefficient D are m^2/s, i.e., D has the same dimensions as the kinematic viscosity ν and the thermal diffusivity a. The difference between D and ν results from the difference between the mass of a diffusing particle and the mass of a molecule of the fluid. The connection between diffusion and viscosity is that both processes involve momentum transfer by collisions.

We note that in the case of a very large pressure gradient, diffusion in a mixture of two materials can result not only from a concentration gradient, but also from the pressure gradient (so-called *barodiffusion*). A pressure gradient in the system can be created by an external field, for example.

We obtain an equation for the concentration c starting from the equation-of-mass balance of the diffusing material. The quantity ρc is the density of diffusing material. The rate of change of this quantity $\partial(\rho c)/\partial t$ is equal to the difference of the fluxes (11.36) through opposing sides of a cube of unit volume, i.e., $-\partial j/\partial x$. Therefore we obtain the one-dimensional diffusion equation for the concentration $c(x,t)$

$$\frac{\partial c}{\partial t} = D \frac{\partial^2 c}{\partial x^2}. \tag{11.37}$$

The generalization of this equation to the three-dimensional case is

$$\frac{\partial c}{\partial t} = D \Delta c. \tag{11.38}$$

This equation is mathematically identical to the heat equation (10.3), where the temperature T is replaced by the concentration c, and the thermal diffusivity a is replaced by the diffusion coefficient D. Therefore we can apply the qualitative results of Sec. 10.1 to diffusion problems by making the appropriate substitutions.

In particular, the linear dimension R of the region occupied by the diffusing material increases in time according to the equation [see (10.6) and also Problem 2 of Sec. 1.2]:

$$R \sim (Dt)^{1/2}. \tag{11.39}$$

Furthermore, letting M be the total mass of the diffusing material, and using the relation

$$M \sim \rho c R^3$$

and (11.39), we obtain an equation for the decrease in concentration with time in the process of diffusion:

$$c \sim \frac{M}{\rho}(Dt)^{-3/2} \tag{11.40}$$

[compare with (10.8)]. Note that if R in (11.39) is the mean-square radius of the region occupied by the diffusing material, then the numerical factor in this formula is $6^{1/2}$.

Using (11.39), we can also easily obtain the typical velocity v of the diffusion of material inside a volume of order R^3:

$$v \sim D/R. \tag{11.41}$$

This velocity must be small in comparison with the random thermal velocities of the particles of the diffusing material between collisions with the particles of the fluid.

Similarly, it is not difficult to obtain estimates for the above quantities in the cases of one-dimensional and two-dimensional diffusion.

We consider diffusion in a fluid of large (macroscopic) particles whose typical linear dimension R is large in comparison with the size of a molecule of the fluid. We estimate the diffusion coefficient in this case.

In the case of diffusion of macroscopic particles, the viscous drag force $F \sim \eta u R$ (the Stokes force), given by (7.20) acts on each particle. Here, u is the velocity of the particle and η is the dynamical viscosity of the fluid.

As above, let c be the mass concentration of the diffusing particles. The density of the material making up the particles themselves is assumed to be equal to the density ρ of the fluid in order to avoid buoyancy forces or the precipitation of the particles. In the case of a gas medium, the density ρ_0 of the diffusing material will obviously be large in comparison with the gas density. In this case we assume that the particles are small enough so that we can neglect the force of gravity during the time of observation of the diffusion process. The quantity $j = \rho c u$ is the mass flux density of the diffusing particles. It can be expressed through the Stokes force F:

$$j \sim \rho c F/(\eta R). \tag{11.42}$$

Equating (11.36) and (11.42), we obtain an estimate for the diffusion coefficient D

$$D \sim \frac{\rho c F}{\rho \eta R}\frac{dx}{dc} \sim \frac{cF}{\eta R}\frac{dx}{dc}. \tag{11.43}$$

Here, ρ is the density of the fluid.

The quantity $dA = F\,dx$ is the work done by the force F over a distance dx. Therefore we obtain from (11.43)

$$D \sim \frac{c}{\eta R}\frac{dA}{dc}. \tag{11.44}$$

The work done by the Stokes force leads to a decrease in the thermal energy of the diffusing particles. In thermodynamic equilibrium the thermal energy per diffusing particle is of the order of the temperature of the fluid T (according to the equipartition theorem). If dc is the change in the mass concentration of diffusing particles, then $\rho\, dc$ is the change in the mass of the diffusing particles per unit volume. Let M be the mass of a diffusing particle. Then $\rho\, dc/M$ is the change in the number of diffusing particles per unit volume. Finally, $T\rho\, dc/M$ is the change in the thermal energy of the diffusing particles per unit volume.

As noted above, this change in thermal energy is equal to the work done by the Stokes forces. The work done per particle is dA, and therefore for $\rho c/M$ particles the total work done is equal to $\rho c\, dA/M$. Hence, from conservation of energy we have

$$T\rho\, dc/M \sim \rho c\, dA/M. \tag{11.45}$$

Using (11.44) and (11.45), we obtain an estimate for the diffusion coefficient

$$D \sim \frac{T}{\eta R}. \tag{11.46}$$

This estimate cannot be obtained purely from dimensional considerations, since D and ν have the same dimensions.

For diffusion in a gas we use (1.38) for the dynamical viscosity η of the gas and therefore we obtain from (11.46)

$$D \sim \frac{\sigma}{R}\left(\frac{T}{m}\right)^{1/2} \sim \frac{\sigma v_t}{R}. \tag{11.47}$$

Here, m is the mass of a gas molecule, σ is the cross section for molecule-molecule collisions in the gas, and $v_t \sim (T/m)^{1/2}$ is the typical thermal velocity of the gas molecules.

In the case of diffusion of heavy molecules in a light gas, an estimate of the diffusion coefficient is [see (1.29)]

$$D' \sim \frac{\langle r\rangle^3}{\sigma} v_t. \tag{11.48}$$

Here, $\langle r\rangle$ is the average distance between molecules of the light gas. The ratio of (11.47) to (11.48) has the form

$$\frac{D}{D'} \sim \frac{\sigma^2}{\langle r\rangle^3 R} \sim \left(\frac{d}{\langle r\rangle}\right)^3 \frac{d}{R} \ll 1. \tag{11.49}$$

Here, we have set $\sigma \sim d^2$, where d is the molecular diameter. Therefore *the diffusion coefficient of large particles is negligibly small in comparison with the diffusion coefficient of a heavy gas in a light gas.* It is not correct to take the limit $R \to d$, since the Stokes force is only applicable when R is much larger than a molecular diameter.

Problem 1. Obtain the estimate

$$\tau \sim \eta R^3/T$$

for the time of rotation of a diffusing macroscopic particle with linear dimension R about its axis. Here, T is the temperature of the fluid and η is its dynamical viscosity.

Problem 2. Show that the diffusion coefficient of sugar in water is of order 1 $\mathrm{cm}^2/\mathrm{day}$ at room temperature.

Problem 3. Assume that the molecules of the fluid jump from point to point in space. Obtain the following dependence of the diffusion coefficient D of a material whose molecules are about the same size as the molecules of the fluid:

$$D \sim \exp(-C/T).$$

Here, C is a constant. Use the estimate (11.39) in working out the solution.

Problem 4. Obtain the estimate (11.46) for the diffusion coefficient of large particles in a fluid using the Einstein relation (1.28).

Chapter 12

Surface phenomena

A given molecule in a liquid is attracted to every other molecule. The net attractive force on a molecule inside the liquid is equal to zero. However, at the surface there is a nonzero net force that tries to decrease the surface area of the liquid to a minimum. The work that must be done to increase the surface area by a small amount dS is

$$dA = \alpha \, dS. \tag{12.1}$$

The quantity α is called the *surface tension.*

Suppose the flat surface of a liquid is transformed into a convex surface with radius of curvature R. The attractive forces between molecules on the surface create an excess pressure directed inward. We estimate this pressure P from dimensional considerations. According to (12.1), the dimensions of surface tension are N/m. The radius of curvature R has units of length. Finally, the dimensions of the pressure P are Pa = N/m^2. Therefore we obtain

$$P \sim \alpha / R \tag{12.2}$$

(Laplace's law). Note that the density ρ of the liquid does not appear in the expression for P because of dimensional considerations.

It follows from (12.2) that the smaller the radius of curvature R the larger the excess pressure. In the case of a spherical drop of radius R, the numerical factor in (12.2) is equal to 2.

As an example of Laplace's law, we estimate the maximum radius of a spherical liquid drop for which the drop will not collapse under its own weight when it lies on a solid surface. The surface-tension force holding the drop together is obtained by multiplying the pressure (12.2) by the surface area of a sphere, i.e., $F \sim \alpha R$. Then the drop will become unstable when $F \sim mg$, where m is the mass of the drop. Substituting $m \sim \rho R^3$, where ρ is the density of the liquid, we find the maximum radius R such that the drop will not collapse under its own weight:

$$R \sim [\alpha / (\rho g)]^{1/2}. \tag{12.3}$$

In particular, for water this radius is about 1 mm.

Problem 1. Show that a rain drop blows up into a bubble and then disintegrates into smaller drops as it falls under gravity in air (at large Reynolds numbers), if the radius of the drop is large in comparison with the estimate (12.3).

Problem 2. Obtain the following qualitative estimate for the surface tension: $\alpha \sim \rho r d$. Here, r is the specific heat of vaporization of the liquid, ρ is its density, and d is the mean interatomic distance.

12.1. Motion of a liquid in a capillary

We estimate the distance to which a liquid rises (or sinks) on the boundary with a solid wall. If the attraction of molecules of the liquid for molecules of the solid is stronger than the attraction of molecules of the liquid for one another, then the wall is "wetted" and the liquid rises on the boundary with the wall; in the opposite case the surface of the liquid sinks near the wall and the wall is not wetted.

Suppose the surface of the liquid rises near the wall. The upward surface-tension force is balanced by the downward gravitational force on the elevated part of the liquid. The excess pressure P due to the elevation of the liquid is of order $\rho g H$, where ρ is the density of the liquid, H is the elevation, and g is the acceleration of gravity. From dimensional considerations the radius of curvature R is the same order of magnitude as the elevation H, since it is impossible to construct a dimensionless quantity from the parameters α, ρ, and g of the problem.

Substituting the above results into Laplace's law (12.2), we find

$$\rho g H \sim \alpha / H,$$

which gives an estimate for H coinciding with (12.3):

$$H \sim [\alpha/(\rho g)]^{1/2}. \tag{12.4}$$

The numerical factor in this formula depends on the degree of wetting of the solid by the liquid.

We next consider a tube of small diameter D placed in a liquid. We estimate the height H to which the liquid rises in the tube above the surface of the liquid. The pressure P formed by the rising column of liquid is equal to $\rho g H$. If the diameter of the tube D is small in comparison with H, then it is evident from geometrical considerations that the radius of curvature R of the surface of the liquid inside the tube is of order D and is independent of H. Substituting the values of P and R into (12.2), we find

$$\rho g H \sim \alpha / D,$$

and therefore

$$H \sim \alpha/(\rho g D). \tag{12.5}$$

This estimate is correct if the height H is large in comparison with the diameter D of the tube, i.e., $\alpha/(\rho g D) \gg D$, from (12.5). We then obtain the condition

$$D \ll [\alpha/(\rho g)]^{1/2}.$$

If this inequality is not satisfied then H decreases to the value (12.4), as one would expect. Hence it follows from (12.5) that the thinner the tube, the higher the liquid rises inside of it.

Problem 1. Show that a surface-active substance which does not dissolve into the liquid, but which is strongly adsorbed to its surface, lowers the surface tension α by a quantity of order nT, where T is the temperature and n is the surface concentration of adsorbed particles (recall that temperature is expressed in energy units).

Problem 2. Bubbles are blown from drops of a solution of a surface-active substance (such as soap) in water with concentration c. Show that the maximum size of a bubble is of order

$$R \sim (cr^3/d)^{1/2},$$

where r is the initial radius of the drop and d is the average intermolecular distance in the liquid.

Problem 3. For the conditions of the preceding problem show that the thickness of the film of the largest possible soap bubble is independent of the size of the initial drop from which it was blown.

Problem 4. Show that an air bubble floating in a liquid oscillates with a period

$$\tau \sim (\rho R^3/\alpha)^{1/2}.$$

Here, R is the radius of the bubble and ρ is the density of the liquid. Show also that the damping time τ' of these oscillations is of order

$$\tau' \sim R\eta/\alpha \gg \tau,$$

where η is the dynamical viscosity of the liquid.

Problem 5. Show that the attractive force between two parallel match sticks of length l floating in water at a separation d from one another is approximately

$$F \sim \alpha^2 l/(\rho g d^2).$$

12.2. Capillary waves

Capillary waves on the surface of a liquid are caused by surface-tension forces. We estimate the dependence of the frequency ω and velocity c_c of capillary waves on the wavelength λ. The frequency ω of capillary waves is determined by the surface tension α, the density of the liquid ρ, and the wavelength λ. Since $[\alpha] = \text{kg/s}^2$, $[\rho] = \text{kg/m}^3$, and $[\lambda] = \text{m}$, it follows from dimensional considerations that

$$\omega \sim \left(\frac{\alpha}{\rho\lambda^3}\right)^{1/2}. \tag{12.6}$$

We see that capillary waves have dispersion, i.e., their velocity

$$c_c \sim \omega\lambda \sim \left(\frac{\alpha}{\rho\lambda}\right)^{1/2} \tag{12.7}$$

depends on the wavelength λ.

Like gravity waves considered in Sec. 6, capillary waves damp out with depth into the liquid over a distance of order λ, which is the only linear dimension in the problem. The wave amplitude (height of the wave crests) also has units of length, but because the equations of motion are linear for small Reynolds number, the wave amplitude appears linearly in the expression for the velocity of the liquid in wave motion. Therefore the argument of the exponential determining the damping of the wave with depth cannot depend on the wave amplitude, since otherwise the dependence of the velocity of the liquid on the wave amplitude would be highly nonlinear.

We consider the conditions under which gravity can be neglected in this problem (when gravity dominates the waves become gravity waves; see Sec. 6.2). The surface energy of the liquid per unit area created by surface tension is by definition equal to α. The kinetic energy of the capillary waves per unit surface area can be estimated from dimensional considerations as

$$E \sim \alpha (a/\lambda)^2. \tag{12.8}$$

Here, a is the oscillation amplitude and λ is the wavelength. The quadratic dependence of E on a is obvious from dimensional considerations, since the velocity of the liquid in the wave is proportional to a. Using the fact that the oscillation energy is proportional to the square of the amplitude of the wave a, for the gravity waves considered in Sec. 6.2 we find that the energy per unit surface area of the liquid is

$$E' \sim \rho g a^2 \tag{12.9}$$

(the dimensions of E and E' are $\mathrm{J/m^2} = \mathrm{kg/s^2}$). The condition for capillary waves is $E \gg E'$. Using (12.8) and (12.9), we have

$$\lambda \ll \left(\frac{\alpha}{\rho g}\right)^{1/2}. \tag{12.10}$$

Therefore *gravity can be neglected for sufficiently short wavelengths* (less than 1 mm for water, for example).

The estimate (12.6) also determines the natural frequency of quadrupole oscillations of a small spherical liquid drop caused by surface-tension forces (spherically symmetric oscillations cannot occur because of the incompressibility of the liquid, while dipole oscillations are forbidden by conservation of the total momentum of the drop). The lowest-frequency oscillations correspond to wavelengths of the order of the radius R of the drop. Replacing λ by R in (12.6), we find

$$\omega \sim \left(\frac{\alpha}{\rho R^3}\right)^{1/2}. \tag{12.11}$$

We note that for the lowest oscillation frequency the numerical factor in this dependence is equal to $2^{3/2}$. According to (12.10), the result (12.11) will be correct when

$$R \ll \left(\frac{\alpha}{\rho g}\right)^{1/2} \tag{12.12}$$

[see also (12.3)].

We next consider the propagation of capillary waves in a shallow liquid. Here, we have the opposite condition from the preceding problem: $H \ll \lambda$, where H is the depth of the liquid. The analogous limiting case for gravity waves was considered in Sec. 6.2. As in the case of gravity waves, the dependence of the square of the frequency ω^2 on H must be linear [see (6.16)]. This linear dependence corresponds to the first term of a Taylor-series expansion of ω^2 in the ratio H/λ. Note that we consider ω^2 rather than ω because the dispersion dependence should not change when the frequency ω changes sign: this follows from the invariance of the equations of motion to a change in the direction of time (neglecting dissipation).

The remaining factors in the dependence of ω^2 on the parameters of the problem can be determined from dimensional considerations. The parameters of interest are the wavelength λ, the density of the liquid ρ, and the surface tension α. Therefore we obtain

$$\omega \sim \frac{1}{\lambda^2}\left(\frac{\alpha H}{\rho}\right)^{1/2}. \tag{12.13}$$

According to (6.15), the velocity of gravity waves in a shallow liquid is independent of the wavelength λ. According to (12.11), on the other hand, the velocity of capillary waves in a shallow liquid

$$c_c \sim \frac{1}{\lambda}\left(\frac{\alpha H}{\rho}\right)^{1/2} \tag{12.14}$$

has dispersion.

We discuss the condition for capillary waves in a shallow liquid. The oscillation energy is determined by the horizontal component of the velocity v_x [see (6.2)]. This component is of order ωa, where a is the wave amplitude. Therefore the energy of the wave per unit surface area of the liquid is roughly

$$E \sim \rho v_x^2 H \sim \rho \omega^2 a^2 H. \tag{12.15}$$

It follows from (12.15) that in order to be able to neglect gravity the frequency of gravity waves (6.16) must be small in comparison with the frequency of capillary waves (12.13), for a given wavelength λ, wave amplitude a, and depth H. The resulting inequality is identical to (12.10). Therefore when (12.10) is satisfied gravity can be neglected for both deep and shallow liquids.

Next, we consider the damping of capillary waves. We first consider the case of a shallow liquid, when $H \ll \lambda$. We use the general expression (7.48) $\gamma \sim \nu/\lambda^2$ for the damping constant of the wave. Here, ν is the kinematic viscosity of the liquid and λ is the wavelength. Substituting λ from (12.13) into this relation, we find

$$\gamma \sim \nu\omega[\rho/(\alpha H)]^{1/2}. \tag{12.16}$$

The condition that the capillary waves not damp out in one or two periods is $\gamma \ll \omega$ [see the discussion after (7.56)]. Using (12.16) we obtain

$$\nu \ll (\alpha H/\rho)^{1/2}. \tag{12.17}$$

When this condition is not satisfied capillary waves are overdamped and the wave process is aperiodic.

We next consider the damping constant of capillary waves in a deep liquid. The wavelength λ in terms of the frequency ω is substituted from (12.6) into the general formula $\gamma \sim \nu/\lambda^2$. We find

$$\gamma \sim \nu(\rho\omega^2/\alpha)^{2/3}. \tag{12.18}$$

In this case it follows from (12.18) that the condition $\gamma \ll \omega$ for capillary waves has the form

$$\nu \ll (\alpha\lambda/\rho)^{1/2}. \tag{12.19}$$

As expected, when $H \sim \lambda$ the inequalities (12.17) and (12.19) become equivalent.

To conclude this section we estimate how capillary waves damp out when the surface of a deep liquid is covered by a thin film. The film vibrates together with the liquid. Treating the film as a solid body, we use the results of Sec. 7.4. The wavelength λ plays the role of the linear dimension of the body. Recall that the wavelength characterizes the thickness of the liquid layer participating in the oscillatory motion of the wave. The surface energy of the liquid per unit area is roughly

$$E \sim \rho v^2 \lambda. \tag{12.20}$$

Here, ρ is the density of the liquid and v is the velocity of the liquid in the wave motion.

Kinetic energy is transformed into heat in a thin surface layer of thickness δ near the film, rather than over the entire thickness λ of the layer. The thickness of the surface layer δ is of order $\delta \sim (\nu/\omega)^{1/2}$ [see (7.41)]. Obviously this statement is correct if $\delta \ll \lambda$, i.e., $\nu \ll \omega\lambda^2$, or, according to (12.6)

$$\nu \ll (\alpha\lambda/\rho)^{1/2}. \tag{12.21}$$

The rate of change of the surface energy of the liquid, using (7.41) for the thickness δ, can be written in the form

$$\dot{E} \sim \frac{\partial}{\partial t}(\rho v^2 \delta) \sim \rho\omega v^2 (\nu/\omega)^{1/2}. \tag{12.22}$$

Dividing (12.22) by (12.20), we obtain an estimate for the damping constant γ of capillary waves in the presence of the film:

$$\gamma \sim (\nu\omega)^{1/2}/\lambda \sim \nu^{1/2}\omega^{7/6}(\rho/\alpha)^{1/3}. \tag{12.23}$$

Here, we have used the estimate (12.6) for the wavelength λ.

The condition $\gamma \ll \omega$ for periodic oscillations again leads to the inequality (12.21), according to (12.23) and (12.6). Therefore we must require the condition (12.21) in order for capillary waves not to damp out after one to two periods. Then the surface layer in which energy is dissipated is always thin.

The effect of the film at the surface of the liquid on the damping of capillary waves is characterized by the ratio of (12.23) to (12.18). We see that this ratio is of order $(\lambda/\delta) \gg 1$. Therefore *when a film is placed on the surface of a liquid the damping of capillary waves increases significantly.*

Problem 1. Show that the natural frequency of multipole oscillations of a liquid drop due to surface-tension forces is proportional to $k^{3/2}$ in the limit $k \gg 1$, where k is the multipole order.

Problem 2. Show that a gas bubble covered by a thin film experiences the same drag in a liquid at small Reynolds numbers as a solid sphere of the same size.

Problem 3. Explain why capillary waves (ripples) appear in front of a body moving in calm water with a small velocity.

Chapter 13

Sound waves

Sound waves in liquids and gases are waves of compression and rarefaction. They arise, for example, in response to small density or pressure disturbances. We consider the general features of the propagation of small disturbances in gases and liquids. In a sound wave the velocity of the fluid is along the direction of propagation of the wave, and therefore sound is a *longitudinal wave*.

The velocity of the fluid in a sound wave is small enough that turbulence does not occur and the term $(\mathbf{v}\cdot\nabla)\mathbf{v}$ in the Navier–Stokes equation (7.3) can be omitted. The condition for neglecting this term is that the Reynolds number must be small:

$$(\mathbf{v}\nabla)\mathbf{v} \ll \partial\mathbf{v}/\partial t.$$

This relation can be written in the approximate form

$$v^2/\lambda \ll \omega v,$$

where λ is the wavelength of the sound wave and ω is the frequency. We then obtain the condition

$$v \ll \omega\lambda \sim c_s. \qquad (13.1)$$

Here, we have used the fact that $c_s \sim \omega\lambda$ is the speed of sound. Therefore *the velocity of the fluid in a sound wave must be very small in comparison with the speed of sound.*

We next consider the condition under which damping of the sound wave can be neglected. According to (7.3), we must have

$$\nu\Delta\mathbf{v} \ll \partial\mathbf{v}/\partial t.$$

This relation can be written in the approximate form

$$\nu v/\lambda^2 \ll \omega v.$$

We then obtain a condition on the kinematic viscosity of the fluid:

$$\nu \ll \omega\lambda^2 \sim c_s\lambda. \qquad (13.2)$$

Neglecting the viscous and nonlinear terms in the Navier–Stokes equation, we obtain the equation

$$\rho\frac{\partial v}{\partial t} = -\frac{\partial P'}{\partial x}. \qquad (13.3)$$

Here, x is the direction of propagation of the sound wave, ρ is the unperturbed density of the fluid, and P' is the difference between the pressure in the wave and the constant uniform pressure P.

In the case of sound waves, the incompressibility equation (6.1) is not applicable. Instead, we obtain the equation of conservation of mass of the fluid in a unit volume. The rate of change of the fluid density $\partial\rho'/\partial t$, where ρ' is the difference between the density in the wave and the unperturbed fluid density ρ, leads to a flux of fluid through the sides of the unit volume. The flux density is ρv, while the difference of the flux densities through opposing sides of the volume perpendicular to the x axis is equal to $\rho\partial v/\partial x$. Here, we have taken the density ρ outside the differentiation sign and have replaced it by the unperturbed density ρ, since inclusion of the density perturbation ρ' leads to the second-order quantity $\rho' v$ [the quantities v, ρ', and P' are all of the same order of magnitude (first order), as will be seen below].

Therefore the mass balance equation has the form

$$\frac{\partial\rho'}{\partial t}=-\rho\frac{\partial v}{\partial x}. \tag{13.4}$$

The small pressure P' and density ρ' perturbations in the sound wave can be connected using the differential relation

$$P'=\frac{\partial P}{\partial\rho}\rho'. \tag{13.5}$$

Here, the derivative $\partial P/\partial\rho$ is calculated for the unperturbed fluid using the equation of state of the fluid.

Substituting (13.5) into (13.4), we eliminate the variable ρ':

$$\frac{\partial P'}{\partial t}=-\rho\frac{\partial P}{\partial\rho}\frac{\partial v}{\partial x}. \tag{13.6}$$

Differentiating this equation with respect to time and substituting dv/dt from the equation of motion (13.3), we obtain a closed equation for the pressure perturbation P' in the sound wave:

$$\frac{\partial^2 P'}{\partial t^2}=\frac{\partial P}{\partial\rho}\frac{\partial^2 P'}{\partial x^2}. \tag{13.7}$$

It is a standard one-dimensional wave equation. Its general solution has the form

$$P'\sim\cos[\omega t-(2\pi x/\lambda)]. \tag{13.8}$$

Substituting (13.8) into (13.7), we find a relation between the frequency ω of the wave and its wavelength λ:

$$\omega=2\pi c_s/\lambda. \tag{13.9}$$

Here, the speed of sound is

$$c_s=\left(\frac{\partial P}{\partial\rho}\right)^{1/2}\sim\left(\frac{P}{\rho}\right)^{1/2}. \tag{13.10}$$

We see from (13.10) that *the speed of sound in liquids is much larger than in gases,* because a much larger force is required to change the density of a liquid by a given fraction.

The solution for the quantities $v(x,t)$ and $\rho'(x,t)$ has the same form (13.8). Furthermore, it is evident from (13.10) that the speed of sound does not depend on the wavelength, i.e., *sound waves do not have dispersion.* We note that the relation (13.9) was already used in the derivation of the inequality (13.2).

The motion of the fluid in a sound wave is a fast enough process that heat generated in the compressed regions of the wave cannot be transferred to the neighboring regions by means of conduction. Hence *sound propagation is an adiabatic process.*

Problem 1. A wave packet of sound waves occupies a finite region of space at each instant of time and propagates in an infinite fluid. Show that the propagation of the wave packet is accompanied by a nonzero transport of matter when averaged over time. Obtain the following estimate for the average momentum of the wave packet

$$\langle p\rangle \sim P'vR^3/c_s^2,$$

where R is the linear dimension of the wave packet.

13.1. Speed of sound

We estimate the speed of sound in a medium and discuss the role of the speed of sound in the flow of a gas around a body.

We first relate the quantities v, ρ', and P' qualitatively. We obtain from (13.4)

$$\omega\rho' \sim \rho v/\lambda.$$

Hence with the help of (13.9) we find

$$v \sim c_s\rho'/\rho. \tag{13.11}$$

Then, substituting (13.5) into (13.11), we find

$$v \sim P'/(\rho c_s). \tag{13.12}$$

We consider the following question: How large can the stream velocity be such that a gas flowing around a solid body can be assumed incompressible? When a gas flows around a body acoustic disturbances are generated which propagate away from the body. The density perturbation ρ' in the sound wave can be related to the pressure perturbation P' using (13.5) and (13.10)

$$\rho' \sim P'/c_s^2. \tag{13.13}$$

Here, c_s is the speed of sound in the gas.

Let u be the velocity of the stream of gas impinging on the body. u is large in comparison with the velocity v of the fluid in the sound wave (we will see this below when we obtain a relation between the velocities u and v). According to (8.4) we have $P' \sim \rho u^2$. Here, we assume that the Reynolds number is large in comparison with unity (see below). Substituting this estimate into (13.13), we finally obtain

$$\frac{\rho'}{\rho} \sim \left(\frac{u}{c_s}\right)^2. \tag{13.14}$$

Therefore *when a gas flows around a body the gas can be considered incompressible if the stream velocity is small in comparison with the speed of sound*: $u \ll c_s$. The ratio $Ma = u/c_s$ is called the *Mach number.*

The expression (13.14) is also correct for liquids. The liquid can be considered incompressible when the stream velocity is small in comparison with the speed of sound in the liquid.

The velocity v of the fluid in the sound waves created by an obstacle is found with the help of (13.11) and (13.14):

$$v \sim \frac{u}{c_s} u \ll u, \tag{13.15}$$

i.e., the velocity of the fluid in a sound wave is very small in comparison with the stream velocity u if $u \ll c_s$.

We estimate the speed of sound in an ideal gas. The equation of state of an ideal gas is

$$MP = \rho T. \tag{13.16}$$

Here, M is the mass of a gas molecule (recall that the Boltzmann constant $k = 1$ and the temperature T is expressed in energy units). From (13.10) and (13.16) we obtain an estimate for the speed of sound in an ideal gas

$$c_s \sim (T/M)^{1/2}. \tag{13.17}$$

In particular, the speed of sound in air is

$$c_s = 20.1 T^{1/2} \text{ m/s}$$

(here the temperature is in degrees Kelvin).

The right-hand side of (13.17) is the thermal velocity of the gas molecules v_t. This follows from the equipartition of energy over the degrees of freedom of the molecules. Hence we conclude that *the speed of sound in a gas is of the order of the thermal velocity of the molecules of the gas.*

As an example, we estimate the velocity of gas flowing into a vacuum from a hole in a container. If the hole is small, the molecules leave the container with the same velocity that they had inside the container, i.e., the thermal velocity $v_t \sim c_s$. Hence we conclude that *a gas flows into a vacuum from a hole with a velocity of the order of the speed of sound of the gas.*

We estimate the Reynolds number for gas velocities of the order of the speed of sound c_s. We have

$$\mathrm{Re} = c_s R/\nu,$$

where R is the linear dimension of the body. From (1.38) the kinematic viscosity ν is of order $c_s l$, where l is the mean free path of the gas molecules. Therefore $\mathrm{Re} \sim R/l$. Because we always require that $R \gg l$ in gas dynamics, it follows that the Reynolds number is always large for gas velocities of the order of the speed of sound. This explains the use of the estimate (8.4) in obtaining (13.14).

Problem 1. Explain why the speed of sound in a liquid increases with increasing pressure.

Problem 2. Show that the speed of sound drops suddenly when gas bubbles are formed in a liquid (*cavitation*).

Problem 3. Consider a gas heated to such a high temperature that the pressure of equilibrium blackbody radiation in the gas is large in comparison with the pressure of the gas itself. Show that the speed of sound is proportional to the square of the temperature.

Problem 4. Estimate the correction to the speed of sound given by (13.17) for a nonideal gas described by the van der Waals equation.

13.2. Propagation of acoustic vibrations

We consider the propagation of sound from an oscillating point source and study how the velocity v of the fluid in the sound wave falls off with the distance r from the source.

Let Q be the acoustic energy radiated by the source per unit time (the power). The power passing through the surface of a sphere of arbitrary radius r surrounding the source must be equal to Q. Let ρ be the density of the gas, and let v be the velocity of the gas in the sound wave. Then ρv^2 is roughly the energy density of the sound wave. The energy flux density is roughly $\rho v^2 c_s$, where c_s is the speed of sound. Multiplying the energy flux density by the surface area of the sphere, we obtain

$$Q \sim \rho v^2 c_s r^2,$$

and hence we have for the velocity of the gas in the sound wave

$$v \sim \left(\frac{Q}{\rho c_s}\right)^{1/2} \frac{1}{r}. \tag{13.18}$$

The velocity of the gas in a sound wave is therefore inversely proportional to the distance from the source. This decrease in the velocity with distance is purely geometrical in origin and is not connected in any way with the attenuation of sound due to viscosity or heat conduction. Under ordinary conditions this "geometric" attenuation of sound is much stronger than attenuation due to dissipative processes in the medium.

Suppose the medium is a cavity with a characteristic linear dimension R filled with a fluid. Stationary acoustic vibrations in the cavity cannot have arbitrary values of the frequency and wavelength: rather, these quantities take on only discrete values. The discrete frequencies are called the natural frequencies of vibration, or the *eigenfrequencies*. We estimate the natural frequencies of vibration of sound waves in a cavity. For the smallest possible frequencies the wavelength is of the order of the cavity size R. It follows from (13.9) that the frequency of these vibrations is of order

$$\omega \sim c_s/R. \tag{13.19}$$

For example, in the case of a spherical cavity of radius R the numerical factor in this formula is equal to 4.5. The higher frequencies are determined from the condition that an integral number of wavelengths must be contained within a distance of the order of the radius of the sphere R. The boundary conditions are that the velocity of the gas must vanish at the surface of the sphere and at the center. Therefore the frequency spectrum is obviously discrete. *The larger the number of different dimensions characterizing the cavity, the denser the frequency spectrum of its acoustic vibrations.*

Problem 1. Show that the velocity of the gas in a sound wave propagating upward in the atmosphere varies with height according to the equation

$$v \sim z^{-1} \exp[Mgz/(2T)],$$

i.e., it first decreases and then increases. Here, M is the mass of a gas molecule, z is the height above the surface of the Earth, g is the acceleration of gravity, and T is the temperature of the gas (assumed to be independent of the height).

Problem 2. A thin tube of diameter d and length l is attached to a resonant cavity with linear dimension R. Show that together with the frequency (13.19) there is a much lower natural frequency

$$\omega \sim c_s d/(lR^3)^{1/2},$$

at which gas enters and leaves the cavity.

Problem 3. A liquid drop of mass m and radius R strikes a fixed barrier with velocity u. Show that the force of impact F is roughly $F \sim mc_s u/R$, where c_s is the speed of sound in the liquid.

13.3. Radiation of sound from a vibrating body

We estimate the acoustic power radiated by a solid body of linear dimension R which vibrates in a fluid with frequency ω and translational velocity amplitude u.

Suppose first that the wavelength λ of the acoustic radiation is small in comparison with the dimensions of the body, i.e., $\lambda \ll R$ or $(\omega R/c_s) \gg 1$. Then we can assume that a given small part of the body emits plane waves independently of the other parts of the body. The velocity of the fluid near the body is equal to the velocity u of the body; therefore, the energy flux density in the plane wave is $\rho u^2 c_s$, where ρ is the density of the fluid and c_s is the speed of sound.

Because the radiation of sound by the different parts of the body is independent, the power of the radiated sound waves can be obtained by multiplying the energy flux density by the surface area of the body, which is of order R^2

$$I \sim \rho c_s u^2 R^2. \qquad (13.20)$$

We see that *for a given velocity amplitude u, the radiated power does not depend on the vibration frequency in the limit of short wavelength.* However, note that for a given vibration amplitude of the body a, the power does depend on the frequency, since $a \sim u/\omega$.

We next consider the opposite limit of long wavelength $\lambda \gg R$, when $(\omega R/c_s) \ll 1$. In this case the intensities emitted by the different parts of the body cannot simply be added together because the sound waves interfere with one another.

We estimate the velocity of the fluid at a distance r from the vibrating body. First, consider distances satisfying $R \ll r \ll \lambda$. Because $r \ll \lambda$, the left-hand side of the wave equation (13.7) can be neglected in comparison with the right-hand side: they differ by the factor $(\lambda/r)^2 \gg 1$. The right-hand side of (13.7) contains the partial derivative $\partial^2 P'/\partial x^2$. In the general three-dimensional case this derivative is obviously replaced by the Laplacian $\Delta P'$. Hence for distances $r \ll \lambda$ the wave equation for P' (and therefore for ρ and $\mathbf{v}$ as well) reduces to Laplace's equation:

$$\Delta \mathbf{v} = 0.$$

We consider the solution of this equation. It was already obtained in Sec. 6.1 [see (6.5)]:

$$\mathbf{v} \sim (\mathbf{u}\nabla)\nabla \left(\frac{R^3}{r}\right). \tag{13.21}$$

Therefore for these distances the motion of the body can be treated as quasisteady. Note that the condition $r \gg R$ implies that the higher order derivatives with respect to r in the solution of Laplace's equation can be neglected; they fall off with r more strongly than the solution (13.21), and therefore their contribution to the velocity $\mathbf{v}$ is smaller by different powers of the small parameter $(R/r) \ll 1$.

Writing the velocity $\mathbf{u}$ as the product of a velocity amplitude and a time dependence of the form

$$f(t) \sim \cos(\omega t + \varphi),$$

we replace the quantity $\mathbf{u}$ in (13.21) by $\mathbf{u} f(t)$, where now $\mathbf{u}$ is the constant velocity amplitude of the body. We then obtain

$$\mathbf{v} \sim (\mathbf{u}\nabla)\nabla \frac{R^3 f(t)}{r}. \tag{13.22}$$

We turn now to distances $r \gtrsim \lambda$. Then the left-hand side of (13.7) cannot be neglected in comparison with the right-hand side: both terms are of the same order of magnitude. The left-hand side of (13.7) can be taken into account by replacing the time t by the retarded time $t - r/c_s$ [see (13.8)]. Then, generalizing (13.22), we obtain

$$\mathbf{v} \sim (\mathbf{u}\nabla)\nabla \frac{R^3 f(t - r/c_s)}{r}. \tag{13.23}$$

Note that the function $f(t - r/c_s)$ appears under the double-differentiation sign $(\mathbf{u}\cdot\nabla)\nabla$ in (13.23). A solution of the form

$$\mathbf{v} \sim f(t - r/c_s)(\mathbf{u}\nabla)\nabla(R^3/r)$$

would be incorrect, since when it is substituted in the wave equation

$$\frac{\partial^2 \mathbf{v}}{\partial t^2} = c_s^2 \Delta \mathbf{v}, \tag{13.24}$$

[generalizing (13.7) to the three-dimensional case], an additional term arises of the form

$$\nabla f \nabla (\mathbf{u}\nabla)\nabla (R^3/r),$$

which is nonzero.

We estimate the expression (13.23) when $r \gg \lambda$. In this case differentiation of the factor $1/r$ with respect to r gives the factor $1/r^2$, while differentiation of the oscillating function $f(t - r/c_s)$ with respect to r gives the factor

$$\frac{\partial f}{\partial r} \sim \frac{1}{c_s}\frac{\partial f}{\partial t} \sim \frac{\omega f}{c_s}.$$

Therefore when $r \gg \lambda$ it is possible to treat the factor $1/r$ as a constant and differentiate only the function f oscillating with frequency ω. Therefore for $r \gg \lambda$ we find from (13.23)

$$v \sim uR^3 \left(\frac{\omega}{c_s}\right)^2 \frac{1}{r}. \tag{13.25}$$

We see that *at large distances* $r \gg \lambda$ (in the so-called *wave zone*) *the velocity of the medium is inversely proportional to the distance from the source of acoustic vibrations.*

We estimate the energy of acoustic vibrations passing through a sphere of radius $r \gg \lambda$ per unit time (i.e., the radiated power):

$$I \sim \rho c_s v^2 r^2 \tag{13.26}$$

[compare with (13.20)]. Substituting (13.25) into (13.26), we see that the arbitrary radius r cancels, as it must. This corresponds to conservation of energy of the acoustic radiation in the absence of dissipation. Finally, we have

$$I \sim \rho c_s u^2 R^2 \left(\frac{\omega R}{c_s}\right)^4. \tag{13.27}$$

The expression (13.27) differs from (13.20) by the factor $(R/\lambda)^4 \ll 1$. Therefore *in the low-frequency case the radiated power is much less than in the high-frequency case.* In the intermediate case $(R/\lambda) \sim 1$ the expressions (13.20) and (13.27) become identical, as one would expect.

We note that for a sphere of radius R whose velocity varies as $\mathbf{u}\cos\omega t$, the numerical factor in (13.27) is equal to $\pi/12$.

Problem 1. A solid body with linear dimension R begins moving in a fluid with velocity u. Obtain the estimate

$$E \sim \rho u^2 R^3$$

for the total energy of acoustic waves emitted by the body. Here, ρ is the density of the fluid.

Problem 2. A cylinder of radius R oscillates harmonically with frequency ω in the direction perpendicular to its axis. Show that the radiated acoustic power per unit length of the cylinder is of order

$$I \sim \rho \omega^3 R^4 u^2 / c_s^2,$$

where ρ is the density of the fluid, c_s is the speed of sound in the fluid, and u is the amplitude of the velocity vibrations of the cylinder.

13.4. Radiation of sound from a pulsating body

Having considered the case where a body radiates sound when it vibrates translationally, we now consider spherically symmetric pulsations of the body, which produce a periodic change in the volume of the body. We estimate the power radiated as sound waves in this case.

Let R be the linear dimension of the body, u its typical velocity of pulsation, and ω the frequency of pulsation. We first consider the radiated power in the case of long wavelength, when $\lambda \gg R$, where λ is the wavelength of the sound wave. Let the distance r lie within the interval $R \ll r \ll \lambda$. As we saw in the solution of the problem of the preceding section, at such distances retardation can be neglected. We estimate the velocity v of the fluid at such distances. The mass flux of fluid displaced by the expanding body per unit time is of order $\rho u R^2$. An equal mass flux must pass through a sphere of radius $r \gg R$, since we are neglecting retardation and the compressibility of the fluid. Estimating this flux as $\rho v r^2$, we obtain

$$\rho u R^2 \sim \rho v r^2,$$

hence we obtain an estimate for the velocity v

$$v \sim \left(\frac{R}{r}\right)^2 u. \qquad (13.28)$$

Recall that this expression is correct for $R \ll r \ll \lambda$.

Equation (13.28) can also be obtained from Laplace's equation $\Delta \mathbf{v} = 0$. Note that a suitable solution of this equation is $\nabla(1/r)$ and not $1/r$, because the solution must fall off with distance as $1/r^2$ (see Sec. 6.1). The solution can be written in vector form

$$\mathbf{v} \sim R^2 u \nabla \frac{f(t)}{r}. \qquad (13.29)$$

The velocity $\mathbf{v}$ is directed along the position vector $\mathbf{r}$. As before, in (13.29) the function $f(t)$ characterizes the variation of the velocity u of the body with time in the pulsational motion of the body.

At distances $r > \lambda$ we must take into account retardation. We then obtain in place of (13.29)

$$\mathbf{v} \sim R^2 u \nabla \frac{f(t - r/c_s)}{r}. \qquad (13.30)$$

In the wave zone, we differentiate f with respect to r in (13.30) and find

$$v \sim \frac{R^2 u \omega}{c_s r}, \quad r \gg \lambda. \qquad (13.31)$$

Substituting (13.31) into (13.26), we obtain the radiated power in the limit of long wavelength:

$$I \sim \rho c_s u^2 R^2 \left(\frac{\omega R}{c_s}\right)^2. \tag{13.32}$$

In the case of a pulsating sphere of radius R, the numerical factor in (13.32) is equal to 2π.

For short wavelengths $\lambda \sim R$ the power (13.32) becomes of order (13.30), as expected, and remains of this order when $\lambda \ll R$.

We compare the results of the present and preceding sections in the case of long wavelength. Comparing (13.27) and (13.32), we see that a body whose volume pulsates with velocity u and frequency ω radiates sound $(\lambda/R)^2$ times more strongly than if it vibrates translationally with the same velocity and frequency. Physically, this is because a change in the volume of the body results in a larger perturbation on the surrounding medium than vibrational motion without a change in volume.

Problem 1. An infinite cylinder of radius R pulsates harmonically with small frequency ω and vibrational velocity amplitude u. Show that the radiated acoustic power per unit length of the cylinder is of order

$$I \sim \rho \omega R^2 u^2.$$

Problem 2. Show that the radiated power per unit mass of a fluid in which there are turbulent velocity pulsations is of order

$$I \sim u^8/(c_s^5 l).$$

Here, u is the typical velocity of turbulent motion, l is its scale of length, and c_s is the speed of sound in the fluid.

13.5. Scattering of sound by obstacles

We estimate the total cross section for scattering of sound waves with frequency ω by a body with linear dimension R. The scattering cross section is defined as the ratio of the power scattered by the body to the flux density of the incident wave. The latter quantity is of order $\rho v^2 c_s$, where c_s is the speed of sound, ρ is the density of the medium (gas or liquid), and v is the velocity of the fluid in the wave near the body.

The wave is scattered as a result of distortion of the incident wave fronts by the body: on the surface of the body the velocity of the fluid must vanish rather than be equal to a given value v. Hence we can treat the scattered wave as a wave created by a body moving with velocity $-v$. Because the velocity v oscillates with frequency ω, the equivalent body also oscillates with the same frequency.

A translationally vibrating body with frequency ω and velocity v radiates sound waves. The radiated power was estimated in Sec. 13.3. This power now becomes the power of the scattered sound waves.

We first consider the case of low frequency, where $\lambda \gg R$. Then from (13.27) the radiated power is of order

$$I \sim \rho c_s v^2 R^2 \left(\frac{\omega R}{c_s}\right)^4. \tag{13.33}$$

Once again we emphasize that the body does not really vibrate under the force of the sound wave with velocity v and frequency ω and thereby radiate: the body is actually at rest!

Dividing (13.33) by $\rho v^2 c_s$, we find the cross section:

$$\sigma \sim \left(\frac{\omega R}{c_s}\right)^4 R^2 \sim \left(\frac{R}{\lambda}\right)^4 R^2. \tag{13.34}$$

We note that for a sphere of radius R the numerical factor in this dependence is equal to $7\pi/9$.

We see that when $\lambda \gg R$ the scattering cross section σ given by (13.34) is small in comparison with the geometric cross section of the body, which is of order R^2. This is because the waves are more or less free to bend around the obstacle, i.e., the phenomenon of diffraction. The cross section (13.34) becomes equal to the geometric cross section when $\lambda \sim R$ and remains constant for $\lambda \ll R$ because of the relation (13.20).

We estimate the force with which a sound wave acts on the body. According to Newton's second law, the force is equal to the momentum transferred from the sound wave to the body per unit time. This momentum is then carried off by the scattered acoustic radiation. The scattered power, according to (13.33) and (13.34), is of order

$$I \sim \rho c_s v^2 \sigma. \tag{13.35}$$

The momentum carried off by the scattered wave per unit time is obtained by dividing the power I carried off by the scattered wave by the velocity of the wave c_s. Hence we obtain for the force F exerted by the sound wave on the body (when $\lambda \gg R$)

$$F \sim \frac{I}{c_s} \sim \rho v^2 \sigma \sim (\omega R/c_s)^4 R^2 \rho v^2. \tag{13.36}$$

We note that for a sphere of radius R the numerical factor in (13.36) is equal to $11\pi/18$.

The acoustic pressure P is obtained from (13.36) by dividing this force F by the surface area of the body $\sim R^2$:

$$P \sim \left(\frac{\omega R}{c_s}\right)^4 \rho v^2 \ll \rho v^2. \tag{13.37}$$

Problem 1. Show that the scattered intensity of sound waves by a body reaches a maximum in the direction opposite to the direction of incidence of the wave.

Problem 2. Explain why the results for the scattering cross section obtained in this section are valid when size of the body R is large in comparison with the amplitude v/ω of the displacement of the fluid in the wave.

Problem 3. Show that in the limit when the wavelength of sound is small in comparison with the dimensions of the body ($\lambda \ll R$) the intensity of the reflected sound is of the same order of magnitude as the intensity of sound scattered over a very small angle of order λ/R.

13.6. Scattering of sound waves by small particles

We estimate the cross section for scattering of sound waves with frequency ω and wavelength λ by small particles with linear dimension $R \ll (\nu/\omega)^{1/2}$. Here ν is the kinematic viscosity of the fluid. It is assumed to be small, such that $(\nu/\omega)^{1/2} \ll \lambda$.

Recall that from (7.41) the quantity $\delta \sim (\nu/\omega)^{1/2}$ determines the thickness of the layer near the body in which the viscosity of the fluid cannot be neglected. In the preceding sections of this chapter we tacitly assumed that $\delta \lesssim R$ and so the thickness δ did not affect the scattering of sound. In particular, when $R \sim \lambda$, this inequality reduces to (13.2).

To determine the cross section for scattering of sound waves by small particles it is convenient first to consider the radiated acoustic power by such a particle when it vibrates in the medium with frequency ω. A similar method was used in Sec. 13.5. The power is determined by (13.26) for $r \gg \lambda$, i.e., in the wave zone. Hence we need to estimate the velocity v of the fluid in the sound wave for $r \gg \lambda$.

The method of solving this problem is to determine v successively, starting with small distances r and matching the solution with the general analytical expressions for v at large distances. Let $u(t)$ be the vibrational velocity of the body. For distances $r \sim R$ we have $v \sim u(t)$.

We next consider distances $R \ll r \ll \delta$. As we saw in estimating the thickness δ in Sec. 7.4, the term $\partial \mathbf{v}/\partial t$ in the equation of motion (7.40) in the case $r \sim \delta$ is of the same order as $\nu \Delta \mathbf{v}$. If $r \ll \delta$, then the term $\partial \mathbf{v}/\partial t$ can be neglected. At the same time, the term $-\nabla P$ can also be neglected in the Navier–Stokes equation (7.3), since, according to (13.3), it is of the same order of magnitude as the term $\rho\, \partial \mathbf{v}/\partial t$. Hence the Navier–Stokes equation (7.3) or (7.40) takes the form of Laplace's equation $\Delta \mathbf{v} = 0$ for distances $r \ll \delta$. The solution of this equation can be written in the form

$$\mathbf{v} \sim \frac{R}{r} \mathbf{u}(t). \tag{13.38}$$

It satisfies the correct boundary condition $v \sim u$ at $r \sim R$. In addition, we know that $\Delta(1/r) = 0$. Since $r \gg R$ we can neglect the other solutions of Laplace's equation, which involve terms of higher order in the ratio R/r.

At first glance it is not clear why we cannot use the result (6.6) and write

$$\mathbf{v} \sim (R/r)^3 \mathbf{u}.$$

Following the discussion of Laplace's equation (6.4), we concluded that a solution of the form (13.38) was unacceptable because it leads to an infinite quantity of fluid carried off by the moving body. However, in the present problem divergences do not occur because the distance r is bounded from above by the quantity δ. Furthermore, over the distance δ the viscosity of the fluid must be taken into account and the solution takes on a different form, which will be discussed below.

Starting from (13.38), we can estimate the order of magnitude of the velocity v at distances $r \sim \delta$, over which the viscosity of the fluid becomes significant:

$$\mathbf{v} \sim \frac{R}{\delta} \mathbf{u} f(t). \tag{13.39}$$

The condition $r \ll \lambda$ implies that retardation can be neglected in (13.39) [see the method of obtaining (13.22)].

Retardation (the term involving $\partial/\partial t$ in the hydrodynamic equations) can also be neglected for distances $\delta \ll r \ll \lambda$. It follows from Sec. 7.4 that for such distances the term $\eta \Delta \mathbf{v}$ involving the viscosity can be neglected in the Navier–Stokes equation (7.3). In particular, if $\delta < R$ the term involving the viscosity can be neglected at all distances r; this fact was used in the preceding sections in discussing the radiation of sound waves.

Hence when $r \gg \delta$ we have potential flow of an ideal fluid. In this case the velocity $\mathbf{v}$ has the functional form (13.22) with an undetermined numerical factor A:

$$\mathbf{v} \sim A(\mathbf{u}\nabla)\nabla \frac{R^3 f(t)}{r}. \tag{13.40}$$

This factor is found from the condition that (13.40) should reduce to (13.39) when $r \sim \delta$. This condition implies that

$$Au \left(\frac{R}{\delta}\right)^3 \sim u \frac{R}{\delta}, \quad \text{or } A \sim \left(\frac{\delta}{R}\right)^2 \gg 1.$$

Therefore (13.40) takes the form

$$\mathbf{v} \sim \frac{\nu R}{\omega} (\mathbf{u}\nabla)\nabla \frac{f(t)}{r}. \tag{13.41}$$

We next consider distances $r \gtrsim \lambda$. In this case retardation in the hydrodynamic equations becomes important. In analogy with the transformation from (13.22) to (13.23), we replace the time t in (13.41) by the retarded time $t - r/c_s$ and obtain

$$\mathbf{v} \sim \frac{\nu R}{\omega} (\mathbf{u} \cdot \nabla)\nabla \frac{f(t - r/c_s)}{r}. \tag{13.42}$$

Suppose, finally, that $r \gg \lambda$. Then (13.42) can be simplified because we only need to differentiate $f(t - r/c_s)$, and not $1/r$. We find

$$v \sim \frac{\nu R u \omega}{c_s^2 r}. \tag{13.43}$$

Hence we have obtained the required estimate of the velocity of the fluid in the wave zone. Note that it is inversely proportional to the distance r from the radiator, as it must be. Substituting (13.43) into (13.26), we find the power I radiated by a small particle:

$$I \sim \rho \nu^2 R^2 \omega^2 u^2 / c_s^2. \tag{13.44}$$

We see that it is proportional to the square of the frequency ω of the sound wave. The numerical factor in (13.44) can be obtained from the quantitative solution for the case of a small sphere of radius R and is equal to $3\pi/2$ in this case.

The radiated acoustic power (13.44) is determined by the kinematic viscosity ν of the fluid. An analogous formula, but with ν replaced by a, where a is the thermal diffusivity of the fluid, can be obtained by repeating the above calculations starting from the heat equation (see Chap. 10). If $\nu \sim a$, then the two mechanisms of acoustic radiation are comparable to one another in power. If the Prandtl number $\mathrm{Pr} = \nu/a \gg 1$, then the viscous mechanism of acoustic radiation dominates.

We now consider the scattering of sound by small particles. As in Sec. 13.5, we conclude that (13.44) determines the power of the scattered sound wave. Dividing it by the flux density $\rho c_s u^2$ of the incident wave, we obtain an estimate for the scattering cross section:

$$\sigma \sim (\nu R \omega / c_s^2)^2 \sim (\delta/\lambda)^4 R^2. \tag{13.45}$$

It is small in comparison with the geometric cross section $\sim R^2$. In addition, it is small even in comparison with the cross section (13.34) for ordinary dipole scattering; this follows from the inequalities stated at the beginning of this section.

When $\delta \sim R$ (13.45) reduces to (13.34), as one would expect.

Problem 1. Show that the differential cross section for scattering of sound by small particles due to the thermal conductivity of the fluid is spherically symmetric. Show also that the differential cross section for scattering of sound by small particles due to the viscosity of the fluid is maximum in the direction of propagation of the sound and in the opposite direction.

13.7. Motion of bodies induced by sound

We formulate the conditions under which a body moves when a sound wave is incident upon it. The inertial force of the body is $M\mathbf{a}$, where $\mathbf{a} = d\mathbf{u}/dt$ is the acceleration and $\mathbf{u}$ is the velocity of the fluid in the neighborhood of the body. The quantity $M \sim \rho_0 R^3$ is the mass of the body, ρ_0 is its density, and R is its characteristic linear dimension. The body will not move if the inertial force is large in comparison with the force acting on it due to the sound wave. Here, we are obviously concerned with small particles and therefore the force produced on the body from the sound waves is the Stokes force $F \sim \eta u R$ caused by the viscous flow of the the fluid around the body (recall the results of the preceding section).

We estimate the inertial force as $M\omega u$, where ω is the frequency of the sound wave. Then the body will not move if the Stokes force is small in comparison with the inertial force:

$$\rho \nu u R \ll \rho_0 R^3 \omega u. \tag{13.46}$$

Hence we obtain a restriction on the size of the body:

$$R \gg \left(\frac{\rho}{\rho_0}\right)^{1/2} \delta \sim \left(\frac{\rho \nu}{\rho_0 \omega}\right)^{1/2}. \tag{13.47}$$

On the other hand, as pointed out in the preceding section, we must have $R \ll \delta$. These inequalities can be satisfied only when $\rho \ll \rho_0$, i.e., in the case when the density of the medium is small in comparison with the density of the body. This condition is not easily satisfied in the case of a liquid. But it is easily satisfied for a gas.

We therefore conclude that the estimate (13.45) for the cross section for scattering of sound by small particles will be correct only if the additional restriction (13.47) is also satisfied. Otherwise, the scattering of sound decreases significantly because the motion of the body induced by the viscous Stokes force is in tandem with the vibrations of the medium.

If $\delta < R$ and the viscosity is small, then we return to the problem considered in Sec. 13.5. In order for the estimate (13.34) of the cross section to be valid, the inertial force must be large in comparison with the acoustic pressure force (13.36):

$$\left(\frac{\omega R}{c_s}\right)^4 R^2 \rho u^2 \ll \rho_0 R^3 \omega u. \tag{13.48}$$

This leads to a restriction on the size of the body:

$$R \gg \left(\frac{R}{\lambda}\right)^4 \frac{\rho}{\rho_0} \frac{u}{\omega}. \tag{13.49}$$

This condition must be satisfied for (13.34) to be valid.

The above discussion pertains to the case of long wavelengths, when $R \ll \lambda$. In the opposite limiting case of short wavelength ($R \gtrsim \lambda$) the acoustic pressure force $\rho u^2 \sigma$ is of order $\rho u^2 R^2$ and in place of (13.48) we will have the inequality

$$\rho u^2 R^2 \ll \rho_0 R^3 \omega u, \tag{13.50}$$

and when this condition is satisfied one can neglect the motion of the body due to the sound wave. From (13.50) we obtain a restriction on the size of the body

$$R \gg \frac{\rho}{\rho_0} \frac{u}{\omega}. \tag{13.51}$$

In the case of a liquid $\rho \sim \rho_0$, and the conditions (13.49) and (13.51) simplify and can be fully satisfied if the amplitude u of the velocity of the liquid in the sound wave is sufficiently small. When $\rho \sim \rho_0$ the condition (13.51) simply states that the amplitude $a \sim u/\omega$ of the vibrations of the fluid in the sound wave must be small in comparison with the size of the scattering body.

13.8. Generation of sound waves by temperature oscillations of a radiator

We estimate the intensity of sound waves in a fluid in contact with a large plane surface whose temperature oscillates with frequency ω and amplitude T'.

According to the results of Sec. 10.1, the temperature oscillations penetrate only a small distance into the fluid and create a thin boundary layer of fluid next to the surface of the body with a thickness x_0 given by (10.11): $x_0 \sim (a/\omega)^{1/2}$. However, the temperature oscillations in the fluid lead to density oscillations

$$\rho' = \frac{\partial \rho}{\partial T} T' = -\rho\beta T'. \tag{13.52}$$

Here, ρ is the unperturbed density of the fluid,

$$\beta = -\frac{1}{\rho}\frac{\partial \rho}{\partial T}$$

is the coefficient of volume expansion of the fluid, T' is the deviation of the temperature of the fluid from the average temperature; the deviation oscillates with the frequency ω.

Oscillations of the density of the fluid lead to oscillations of the velocity v of the fluid. The velocity v can be expressed in terms of ρ' with the help of the mass balance equation (13.4). Using (13.52), this equation can be rewritten in the form

$$\beta \frac{\partial T'}{\partial t} = \frac{\partial v}{\partial x}. \tag{13.53}$$

Equation (13.53) can be written in approximate form as

$$\beta\omega T' \sim v/x_0,$$

and therefore

$$v \sim \beta(a\omega)^{1/2} T'. \tag{13.54}$$

Here, we have used (10.11) for the thickness x_0 of the boundary layer.

Since the propagation of sound waves created by the density oscillations ρ' is one dimensional, geometric damping does not occur and v is also the typical velocity of the fluid outside the boundary layer, as long as we neglect dissipative processes in the sound wave.

We will assume that the wavelength $\lambda \sim c_s/\omega$ is small in comparison with the dimensions of the surface (which was already assumed to be large). In this case we have the limit of short wavelength and the radiated power is described by (13.20), where R^2 is now the area of the surface radiating the sound. The velocity v given by (13.54) plays the role of the velocity u. Dividing (13.20) by R^2 and using (13.54), we obtain the radiated power per unit surface area:

$$I \sim \rho v^2 c_s \sim \rho c_s \beta^2 a\omega T'. \tag{13.55}$$

It follows that *the radiated power is proportional to the first power of the frequency ω of the temperature oscillations of the surface.* In the exact solution of the problem the numerical factor in (13.55) turns out to be $\frac{1}{2}$. These results are valid if the thickness of the boundary layer x_0 is small in comparison with the wavelength of the sound wave λ, i.e., $x_0 \ll \lambda$. Otherwise, we simply have a damped thermal wave. This inequality can be rewritten in the form $\omega \ll c_s^2/a$.

Problem 1. Estimate the power of radiated sound waves from a sphere of radius R whose temperature oscillates with frequency ω. The amplitude of the temperature oscillations is T'.

Problem 2. At time $t=0$ the temperature of a sphere of radius R suddenly changes by an amount T'. Estimate the total energy of the sound waves radiated by the sphere in terms of the thermal diffusivity a of the fluid in which the sphere is found, its density ρ, and the coefficient of volume expansion of the fluid β.

13.9. Sound propagation in pipes

In this section we consider the propagation of sound in a pipe whose diameter is small in comparison with the wavelength of the sound λ.

We first estimate the acoustic power radiated from the open end of a pipe of radius R in which sound waves propagate. We assume long wavelength ($\lambda \gg R$). Sound waves propagate along the axis of the pipe with the velocity

$$c_s \sim (T/M)^{1/2}. \quad (13.56)$$

We have assumed an ideal gas. Because of the sound waves, the gas at the open end of the pipe oscillates periodically. Let u be the typical velocity of the gas in the sound wave. Then we have a gas with a surface area of order R^2 pulsating with velocity u and frequency $\omega \sim c_s/\lambda$. A pulsating body emits sound waves; see Sec. 13.4. The radiated power is given by the estimate (13.32). This formula is qualitatively correct not only for the pulsations of a sphere with a surface area of order R^2, but also for volume pulsations of any body whose pulsating surface has an area of order R^2.

In contrast to volume pulsations of a solid body, a sound wave in a pipe also produces an excess pressure P', which pulsates with the frequency ω. This pressure in turn creates its own sound wave of frequency ω. However, this effect is second order and we will ignore it here.

Hence, according to (13.32), the power of sound radiated from the open end of a pipe of radius R is of order

$$I \sim \left(\frac{R}{\lambda}\right)^2 \rho c_s u^2 R^2, \quad (13.57)$$

where ρ is the density of the gas.

We consider now the decrease of the acoustic energy inside the pipe with time as a result of the radiation of sound according to (13.57). Let l be the length of the pipe. The acoustic energy in the pipe can be estimated as

$$E \sim \rho u^2 R^2 l. \quad (13.58)$$

It obeys the equation

$$dE/dt = -I,$$

where I is defined by (13.57). Writing this equation in the form

$$dE/dt = -E/\tau,$$

we see that τ is the time constant of the exponential decay of the acoustic energy inside the pipe. We obtain from (13.57) and (13.58)

$$\tau = E/I \sim (l/c_s) \cdot (\lambda/R)^2. \quad (13.59)$$

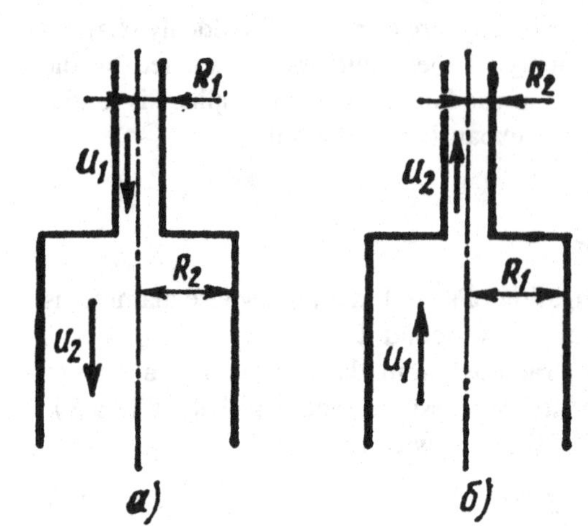

FIG. 21. Radiation of sound from a narrow pipe into a wide one (a) and from a wide pipe into a narrow one (b).

This discussion is correct when the relaxation time τ is large in comparison with the period ω^{-1} of the sound wave (otherwise the sound wave damps out aperiodically). Therefore from the condition $\omega\tau \gg 1$ we obtain

$$l \gg \frac{R}{\lambda} R. \tag{13.60}$$

Because $R \ll \lambda$, the length of the pipe does not have to be large in comparison with the radius R.

We next consider the fraction of acoustic energy transmitted in a pipe when its diameter changes. For simplicity we consider an idealized problem in which it is required to estimate the fraction of acoustic energy transmitted from a pipe of radius R_1 into a pipe of much larger radius $R_2 \gg R_1$ (Fig. 21a). We assume $\lambda \gg R_1$, $\lambda \gg R_2$, i.e., we have the limit of long wavelength in both pipes.

The quantity of gas $\rho u_1 R_1^2$ flowing from pipe 1 into pipe 2 per second (u_1 is the velocity of the gas in the sound wave in pipe 1) must be equal to the quantity of gas $\rho u_2 R_2^2$ flowing into pipe 2 (u_2 is the velocity of the gas in the sound wave propagating in pipe 2). The quantity of gas $\rho u_2 R_1^2$ which goes back into pipe 1 from pipe 2 due to the reflection of the sound wave from pipe 2 can be neglected because of the inequality $R_1 \ll R_2$. Hence we obtain

$$u_1 R_1^2 \sim u_2 R_2^2. \tag{13.61}$$

The fluxes of acoustic energy in pipes 1 and 2 are

$$I_1 \sim \rho c_s u_1^2 R_1^2, \quad I_2 \sim \rho c_s u_2^2 R_2^2.$$

Using (13.61), we obtain for the ratio of the transmitted acoustic energy to the energy incident from pipe 1 into pipe 2

$$\frac{I_2}{I_1} \sim \left(\frac{R_1}{R_2}\right)^2 \ll 1. \tag{13.62}$$

Hence *only a small fraction of the acoustic energy is transmitted from a narrow pipe into a wide one*. The estimate (13.62) is also qualitatively correct for a pipe in the form of a fluted horn, where R_1 is the initial radius of the horn and R_2 is its final radius.

If pipe 2 is so wide that $R_2 \gg \lambda$, then we must use (13.57) and then we have in place of (13.62)

$$\frac{I_2}{I_1} \sim \left(\frac{R_1}{\lambda}\right)^2 \ll 1. \qquad (13.63)$$

Comparing (13.62) and (13.63), we see that *use of a horn with outer radius $R_2 \ll \lambda$ increases the intensity of the emitted sound by a factor of $(\lambda/R_2)^2 \gg 1$.*

Suppose next that sound is incident from a pipe of large radius R_1 into a pipe of much smaller radius $R_2 \ll R_1$ (Fig. 21b). In this case on the boundary between the pipes gas is incident from pipe 1 into pipe 2 as well as from pipe 2 into pipe 1. Hence we have $u_1 = u_2$. The ratio of the fluxes of acoustic energy (i.e., the intensities of the sound waves I_1 and I_2 in the two pipes) is given by

$$\frac{I_2}{I_1} \sim \left(\frac{R_2}{R_1}\right)^2 \ll 1. \qquad (13.64)$$

Hence we conclude that *in both limiting cases $R_2 \ll R_1$ and $R_2 \gg R_1$ only a small fraction of the acoustic energy is transmitted from one pipe to the other.* However, in the intermediate case $R_2 \sim R_1$ the fraction of acoustic energy transmitted from one pipe to the other is not small, as is evident from (13.62) or (13.64).

Problem 1. One end of a cylindrical pipe is open, while the other end is closed by a membrane. Show that when the vibrational frequency of the membrane is equal to a natural frequency of sound in the cylindrical pipe, the velocity of the gas at the open end of the pipe grows without bound as a result of resonance.

13.10. Absorption of sound

We consider sound absorption in a liquid or gas. We first consider the viscous part of the sound absorption. As we have seen above, the flux density of acoustic energy is roughly $W \sim \rho c_s u^2$. Here, ρ is the density of the fluid, u is the velocity of the fluid in the sound wave, and c_s is the speed of sound. According to (7.47), the rate of dissipation of energy into heat per unit volume is

$$j \sim \rho \nu u^2/\lambda^2. \qquad (13.65)$$

Here, ν is the kinematic viscosity of the fluid and λ is the wavelength of the sound wave. The result (7.47) was obtained for gravity waves on the surface of a liquid, but when written in this general form it can be used for any type of oscillation.

The *absorption coefficient of sound* α_s is defined as the ratio of j to W:

$$\alpha_s = \frac{j}{W} \sim \frac{\nu}{c_s \lambda^2} \sim \frac{\nu \omega^2}{c_s^3}. \qquad (13.66)$$

Hence *the absorption coefficient of sound is proportional to the square of the frequency* ω.

The thermal conductivity of the fluid makes a similar contribution to sound absorption, which is given by (13.66) with ν replaced by the thermal diffusivity of the fluid a.

Note the analogy between the absorption coefficient α_s given by (13.66) and the damping coefficient in a medium given by (7.48). In the case of sound these quantities differ only by a constant factor (the speed of sound U).

Because $j = -dW/dx$, where x is the direction of propagation of the sound wave, we obtain from (13.66)

$$\frac{dW}{dx} = -\alpha_s W. \tag{13.67}$$

Therefore the acoustic energy W decreases exponentially with the distance x, and the quantity α_s^{-1} is the effective absorption length of the sound. This length should be large in comparison with the wavelength λ of the sound wave, since otherwise sound propagation would not occur, in general. Therefore (13.66) is applicable when $\alpha_s^{-1} \gg \lambda$ or

$$\nu \ll c_s\lambda \sim c_s^2/\omega. \tag{13.68}$$

Similarly, we obtain

$$a \ll c_s\lambda.$$

This condition restricts the wavelength from below. As expected, it is equivalent to (13.2).

Note that up to now we have assumed that the speed of sound is given by (13.10), which assumes an absence of heat transfer. However, if the thermal conductivity of the fluid is sufficiently large, then the speed of sound (13.10) is defined at constant temperature. Indeed, when the thermal conductivity is large, the temperature of the medium can reach a steady state during the compressions and rarefactions of the sound wave. For an ideal gas the speed of sound remains of the order of magnitude of the thermal velocity of the gas molecules (13.17); however, the numerical factor in this dependence is different from the case when the thermal conductivity is small and the sound is isoentropic.

We estimate the condition when sound is isoentropic and when it is isothermal. We consider the isoentropic case first. The typical time of rarefaction or compression of the gas in a sound wave is of order ω^{-1}. The time required for heat to be transferred a distance of the order of a wavelength λ is, according to (10.7),

$$\tau \sim \lambda^2/a.$$

Therefore the acoustic process will be adiabatic when $\omega\tau \gg 1$ or

$$a \ll c_s\lambda. \tag{13.69}$$

This is equivalent to the existence condition (13.68) for sound waves (i.e., sound absorption must be small).

When the wavelength is small so that $a \sim c_s\lambda$ the absorption coefficient of sound (13.66) is roughly

$$\alpha_s \sim c_s/a. \quad (13.70)$$

Hence $\alpha_s \sim \lambda^{-1}$, and so the absorption of sound over a wavelength λ is significant. Therefore the acoustic sound wave is aperiodic and sound does not really exist.

Next we consider still shorter wavelengths λ, when $a \gg c_s\lambda$ [the opposite condition to (13.69)]. Then sound propagation is an isothermal process. We estimate the absorption coefficient of sound in this limiting case. We write the acoustic energy density in the form $E \sim \rho c_P T$. Here, c_P is the specific heat of the fluid at constant pressure and T is its temperature. The flux density of acoustic energy has the form

$$W \sim c_s E \sim c_s \rho c_P T. \quad (13.71)$$

The rate of change of the energy density of the sound wave (i.e., the power density of the sound wave) is

$$j = \frac{\partial}{\partial t}(\rho c_P T),$$

and according to the heat equation (10.2) it can be written in the form

$$j \sim \rho c_P a T/l^2. \quad (13.72)$$

Here, l is the typical distance over which the temperature of the fluid varies. In this case it is not the wavelength λ, but is much larger because of the equalization of the temperature due to the large thermal conductivity of the fluid. We note that the excess density ρ' and pressure P' in a sound wave vary periodically over the wavelength λ, whereas the temperature varies over the much larger distance l. We estimate this distance.

A flux density of acoustic energy (13.71) is generated by a temperature gradient and, according to (10.1), it is equal to $\rho c_P a\, dT/dx$. Approximating this quantity as $\rho c_P a T/l$ and setting it equal to (13.71), we obtain the distance l:

$$l \sim a/c_s \gg \lambda. \quad (13.73)$$

Dividing (13.72) by (13.71) and using (13.73), we find the absorption coefficient of isothermal sound:

$$\alpha_s \sim a/c_s l^2 \sim c_s a/a \ll \lambda^{-1}. \quad (13.74)$$

We see that the estimates (13.74) and (13.70) coincide. Because of the smallness of the absorption coefficient (absorption is weak over a wavelength λ) *isothermal sound can propagate over large distances*, as in the case of adiabatic sound considered earlier.

We considered sound absorption in a fluid due to transformation of acoustic energy into heat. We turn now to another mechanism of sound absorption in a fluid caused by the presence of small scatterers. When sound waves scatter from small particles acoustic energy is dissipated into heat. We estimate sound absorption due to small particles in a fluid, where

$$R \ll (\nu/\omega)^{1/2} \ll \lambda.$$

Here, λ is the wavelength of the sound wave, ω is its frequency, and ν is the kinematic viscosity of the fluid.

Let u be the typical velocity of the fluid in the sound wave. Dissipation of acoustic energy into heat takes place over a length of the order of the dimension R of the scattering particle. The absorbed power E is given by (7.45):

$$\dot{E} \sim \rho\nu u^2 R.$$

Here, ρ is the density of the fluid. Although the viscosity of the fluid must be taken into account over much larger distances $\delta \sim (\nu/\omega)^{1/2}$ (see Sec. 7.4), energy dissipation over a length of order δ is negligible. Indeed, according to (13.38) and (13.39), the velocity of the fluid at a distance $r \sim \delta$ from the scatterer (the scatterer is a source of secondary sound waves emitted in all directions) is of order

$$v \sim \frac{R}{r} u \sim \frac{R}{\delta} u. \tag{13.75}$$

The power j absorbed per unit volume of fluid is given by (7.43):

$$j \sim \rho\omega v^2.$$

Therefore the power $\dot{E}'$ absorbed over a distance δ from the scatterer is of order

$$\dot{E}' \sim j\delta^3 \sim \rho\omega u^2 R^2 \delta, \tag{13.76}$$

and the ratio of this power with $\dot{E}$ is

$$\frac{\dot{E}'}{\dot{E}} \sim \frac{\omega R\delta}{\nu} \sim \frac{R}{\delta} \ll 1. \tag{13.77}$$

Therefore we have shown that *the dissipation of acoustic energy into heat in the process of scattering of sound by small particles takes place over a distance of the order of the dimension R of the scatterer.*

The absorption cross section σ' is defined as the ratio of the power dissipated in the scattering of the sound wave by the particle to the flux density of the incident acoustic energy $W \sim \rho c_s u^2$. Therefore we find

$$\sigma' = \frac{\dot{E}}{W} \sim \frac{\nu R}{c_s}. \tag{13.78}$$

In contrast to the scattering cross section σ given by (13.45), the absorption cross section σ' is independent of the frequency of the sound wave. We also note that for spherical particles of radius R the proportionality constant in (13.78) is 6π.

The analogous absorption cross section due to the thermal conductivity of the fluid is obtained from (13.78) by replacing ν with the thermal diffusivity a.

We compare the absorption cross section σ' with the scattering cross section (13.45) obtained under the same conditions $R \ll \delta \ll \lambda$:

$$\frac{\sigma'}{\sigma} \sim \frac{c_s^3}{\nu\omega^2 R} \sim \frac{\lambda}{R}\left(\frac{\lambda}{\delta}\right)^2 \gg 1. \tag{13.79}$$

Hence *sound absorption by small particles dominates over scattering by small particles.*

Next, we estimate the absorption coefficient of sound by small particles. Let n be the concentration of scatterers. Then, the expression $n\dot{E}$, where $\dot{E}$ is given by (7.45)

$$\dot{E} \sim \rho \nu u^2 R,$$

is the energy density transformed into heat per unit time. Substituting this estimate into the definition (13.66) for the absorption coefficient α_s and using the fact that the flux of the incident sound wave has the form

$$W \sim \rho c_s u^2,$$

we find the absorption coefficient of sound by small particles:

$$\alpha_s = \frac{j}{W} \sim \frac{n\dot{E}}{W} \sim n\sigma' \sim \frac{n\nu R}{c_s}. \tag{13.80}$$

As mentioned above, the quantity α_s^{-1} characterizes the distance over which significant sound absorption occurs. In order for (13.80) to be applicable, we must have the condition $\alpha_s^{-1} \gg \lambda$, or

$$n \ll \frac{c_s}{\nu R \lambda} \tag{13.81}$$

(otherwise the sound wave would damp out aperiodically).

When (13.81) is satisfied, the length α_s^{-1} is large in comparison with the average distance $n^{-1/3}$ between scatterers. Indeed, the inequality

$$\alpha_s^{-1} \gg n^{-1/3}$$

with the use of (13.80) reduces to

$$n \ll \left(\frac{c_s}{\nu R}\right)^{3/2}. \tag{13.82}$$

We show that when (13.81) is satisfied, (13.82) is also satisfied. That is, we prove that

$$\frac{c_s}{\nu R \lambda} \ll \left(\frac{c_s}{\nu R}\right)^{3/2}. \tag{13.83}$$

Rewriting (13.83) in the form

$$\lambda^3 \gg R\delta^2,$$

we see that this relation is obvious, since $\lambda \gg R$ and $\lambda \gg \delta$.

Therefore the scatterers form a continuous medium from the point of view of sound absorption. However, from (13.81), the concentration of scatterers n cannot be so large that the sound wave does not propagate in the medium. For a given density of scatterers the inequality (13.81) places a restriction on the particle radius R. The particle radius must also satisfy the condition $R \gg \delta \sim (\nu/\omega)^{1/2}$.

Problem 1. Using the estimate (13.66) and the expression (1.38) for the kinematic viscosity ν of a gas, obtain the estimate

$$\alpha_s \sim M\omega^2/(\sigma n_0 T)$$

for the absorption coefficient of sound in a gas. Here, ω is the frequency of the sound wave, M is the mass of a gas molecule, T is the gas temperature, σ is the cross section for molecule-molecule collisions in the gas, and n_0 is the concentration of gas molecules.

Problem 2. Show that the ratio of the absorption coefficient of sound by small particles to the absorption coefficient in the gas itself is of order $nR\lambda^2$, where n is the concentration of small particles, R is the linear dimension of a particle, and λ is the wavelength of the sound wave.

Problem 3. Obtain the estimate

$$\alpha_s \sim (\nu\omega)^{1/2}/(Rc_s)$$

for the absorption coefficient of sound propagating in a pipe of radius R. Here, c_s is the speed of sound, ν is the kinematic viscosity of the gas, and ω is the frequency of the sound wave.

Problem 4. Explain why in liquids sound absorption is determined mainly by viscosity, while the contribution of the thermal conductivity is negligibly small. Also, explain why the absorption coefficient of sound is much smaller in liquids than in gases for the same frequency.

Problem 5. A sound wave is reflected from a solid wall. Show that the fraction of acoustic energy absorbed in the wall is of order

$$(a\omega)^{1/2}/c_s,$$

where a is the thermal diffusivity of the fluid in contact with the wall, c_s is the speed of sound in the fluid, and ω is the frequency of the sound wave. Assume that the sound waves do not penetrate into the wall and that the temperature of the wall is constant because of its large heat capacity.

13.11. Propagation of supersonic disturbances

When a fluid flows around a body at subsonic speeds, the presence of the obstacle changes the motion of the fluid everywhere, both upstream and downstream. The perturbation created by the body propagates in all directions in the fluid with the speed of sound.

New effects arise when the stream velocity is larger than the speed of sound. An example has already been discussed in Sec. 13.1: the compressibility of the liquid or gas must be taken into account when the flow is supersonic. In addition, when the flow around a body is supersonic, the presence of the body affects only the region behind it and immediately in front of it.

We consider a typical example involving the propagation of supersonic disturbances. A large quantity of energy E is released at a point in a gas at the initial time (a point explosion). We consider how the disturbance propagates in space.

When the energy release is large the pressure P of the gas inside the region disturbed by the explosion is large in comparison with the pressure P_0 of the gas in the undisturbed region. Hence the quantity P_0 can be neglected in estimating the solution of the problem. At a given instant of time, the boundary of propagation of the disturbance is a sphere of radius R; it is called a *shock wave*. We assume that its propagation velocity is larger than the speed of sound.

The radius R of the shock wave front is determined by the energy E, the density ρ_0 of the unperturbed gas, and the time t. The quantity E/ρ_0 has the dimensions $\mathrm{J\,m^3/kg = m^5/s^2}$. The quantity Et^2/ρ_0 has the dimensions $\mathrm{m^5}$. Therefore we obtain

$$R \sim (Et^2/\rho_0)^{1/5}. \tag{13.84}$$

Hence *the radius of a strong shock wave expands in time as* $t^{2/5}$. The proportionality constant in (13.84) depends on the medium in which the wave propagates. For air it is equal to 1.03.

The velocity of the front of the shock wave, and also the velocity of the gas behind the front, can be estimated qualitatively as R/t. We obtain from (13.84)

$$u \sim \left(\frac{E}{\rho_0 t^3}\right)^{1/5}. \tag{13.85}$$

Hence the velocity u of the shock wave falls off with time as $t^{-3/5}$. The velocity of the gas behind the front is less than the velocity u of the front because the gas cannot overtake the front.

In the case of a strong explosion the pressure p inside the disturbed region reduces to the kinematic pressure (8.4), i.e., $P \sim \rho u^2$. Here, ρ is the density of the disturbed gas. We will see below that although the density of the disturbed gas is large, it is of the same order of magnitude as the unperturbed gas density ρ_0. Using (13.85), we find

$$P \sim \rho_0 [E/(\rho_0 t^3)]^{2/5}. \tag{13.86}$$

We see that for a fixed density ρ_0 and time t, the perturbed pressure P can be arbitrarily large when the energy of the explosion is sufficiently large. This justifies the above approximation of neglecting the unperturbed pressure P_0. The pressure P falls off with time as $t^{-6/5}$.

The above solution becomes inapplicable for large times, when the pressure P in (13.86) becomes of the order of the unperturbed pressure P_0. This determines the lifetime of the shock wave

$$\tau \sim [E/(\rho_0 c_s^5)]^{1/3}. \tag{13.87}$$

Here, $c_s \sim (P_0/\rho_0)^{1/2}$ is the speed of sound in the gas.

Hence we obtain $P \sim P_0 \sim \rho_0 u^2$. According to (13.10), we then have $u \sim c_s$, i.e., the velocity u of the shock wave becomes equal to the speed of sound. Therefore *the shock wave weakens and finally transforms into an ordinary sound wave.*

Next, we prove that the perturbed density ρ in a strong shock wave is the same order of magnitude as the unperturbed density ρ_0. Let T be the perturbed temperature of the gas inside the perturbed region. Then $\rho c_P T R^3$ is the energy of the perturbed gas. It follows from conservation of energy that this energy must be equal to E:

$$\rho \sim E/(c_P T R^3). \tag{13.88}$$

We estimate the temperature T. When $u > c_s$ the quantity u is the velocity of the gas molecules; hence $T \sim Mu^2$, where M is the mass of a gas molecule. Using (13.85), we find

$$T \sim M\left(\frac{E}{\rho_0 t^3}\right)^{2/5}. \tag{13.89}$$

Therefore *the temperature of the gas inside the perturbed region decreases with time as* $t^{-6/5}$ [the same as the pressure; see (13.86)]. Substituting (13.84) and (13.89) into (13.88) and using the fact that the heat capacity c_P is of order M^{-1} (recall that the Boltzmann constant is taken to be equal to unity), we find $\rho \sim \rho_0$. Finally, numerically we have $\rho > \rho_0$. For example, in the case of a diatomic gas the maximum compression of the gas by the shock wave is $\rho/\rho_0 = 6$.

Hence we have shown that *for an arbitrarily strong shock wave the compression of the gas cannot be arbitrarily strong*: The gas density behind the front of the shock wave is of the same order of magnitude as the density of the unperturbed gas. This statement is true not only for a point explosion, but is a very general property of shock waves.

In the problem considered above we did not specify the mechanism of the energy release at a point in space. This is a serious deficiency of the above discussion, since the results are extremely sensitive to the mechanism. For example, suppose that at the initial time a "spherical piston" begins to expand uniformly into a gas with a large velocity $v > c_s$. We estimate the velocity of the resulting shock wave. The estimate (13.85) does not apply: it is valid only for later times when the piston has stopped expanding. Let u be the velocity of the shock wave. It is determined by the velocity v of the piston, the speed of sound c_s, and possibly by the time t, as in the preceding problem. However, it can be shown from dimensional considerations that u cannot depend on the time t. Then we can write

$$u = c_s f\left(\frac{v}{c_s}\right), \tag{13.90}$$

where f is a function which cannot be determined from dimensional considerations. The speed of sound $c_s \sim (P_0/\rho_0)^{1/2}$ is the combination of the unperturbed pressure P_0 and density ρ_0 on which the velocity u depends.

In the limit $v \ll U$ we have $f \to 1$, and the wave front of the disturbance becomes the front of a sound wave. Expanding the function f in a Taylor series in $(v/c_s) \ll 1$ and keeping only the first term, we find

$$u - c_s \sim v. \tag{13.91}$$

The above result is also valid for a piston in a cylindrical tube compressing the gas in front of it as it moves with a large velocity v. In particular, in the case of air the numerical factor in (13.91) is equal to 0.6 for a cylindrical tube.

In the opposite case $v \gg c_s$ the speed of sound must drop out of (13.90) because transport of the disturbance by sound waves cannot keep up with the rapid motion of the piston. In other words, like the case of a strong explosion considered above, the velocity of a strong shock wave should not depend on the unperturbed pressure P_0 of the gas and therefore it should be independent of the speed of sound, which is related to P_0 by

$$c_s \sim (P_0/\rho_0)^{1/2}.$$

Hence in the limiting case when the piston velocity is large, the function f must be a linear function of v/c_s, and therefore c_s drops out of (13.90), and we have $u \sim v$ when $v \gg c_s$. In the case of a piston in a cylindrical tube compressing the gas with a large velocity $v \gg c_s$, the numerical factor in this dependence is 1.2 for air. The velocity of the shock wave is always larger than the piston, i.e., $u > v$.

Problem 1. Explain why the entropy of a gas increases when the streamlines pass through a shock wave.

Problem 2. Consider the motion of a body in an ideal fluid. Explain why d'Alembert's paradox does not occur when shock waves can propagate in the fluid and show that the drag force on the body is nonzero.

Problem 3. Show that in a weak shock wave the entropy jump across the front of the wave is proportional to the cube of the pressure jump.

Problem 4. A plane sound wave is normally incident on the front of a shock wave. Show that the pressure of the transmitted sound wave is always larger than the pressure of the incident wave.

Problem 5. Show that in a weak shock wave the finite width of the front due to the viscosity of the gas is inversely proportional to the pressure jump.

Problem 6. Show that in a strong shock wave the width of the front is of the order of the mean free path of the gas molecules.

Problem 7. Show that a gas cannot reach supersonic velocities when flowing through a tapered nozzle.

Problem 8. Consider a body moving with a supersonic velocity in a gas. Explain why a shock wave is generated at the front of the body (the so-called *head shock wave*) whose front is stationary with respect to the body. Show that at infinity the surface of the shock wave intersects the direction of motion of the body at an angle (the so-called *Mach angle*):

$$\alpha = \arcsin(c_s/u).$$

Here, c_s is the speed of sound and u is the velocity of the body with respect to the gas.

Problem 9. Explain the fact that in a strong point explosion in a gas almost the entire mass of gas included in the explosion wave is localized within a thin layer at the back surface of the shock-wave front.

Conclusion

The present text contains a large number of different problems in hydrodynamics, gas dynamics, and physical kinetics. At first glance, it would seem that the basic method of solution is dimensional analysis. In actuality, in most of the problems the result depended on a dimensionless combination of the physical parameters and the main problem was to find how this dimensionless quantity figured in the answer. This problem was solved by using physical intuition and the conservation laws, and by analogy with the method of solution of other similar problems.

The method of the *self-similar solution* is widely used in hydrodynamics and gas dynamics. For example, we used this method in the final section of the book in the solution of the problem of a strong point explosion in a gas: We used dimensional considerations and the law of conservation of energy in finding the dependence of all of the interesting physical quantities on time; only the constant factors in the solution remained unknown.

In other problems the solution may not be so detailed. For example, if a strong point explosion occurs in a porous medium with a high degree of porosity rather than in a gas, part of the energy of the explosion is transformed into heat, and part into the kinetic energy of the ground covered by the explosion. In this case dimensional considerations can only determine, for example, the relation between the velocity of the front of the wave and the mass of material enclosed by the shock wave.

In order to determine quantitatively the kinetic transport coefficients in the kinetic theory of gases, an effective approximate method of solving the Boltzmann equation is needed. We avoided this question by replacing the collision integral in the Boltzmann equation by a simple expression involving the concept of mean free path. Obviously, this approach is correct only qualitatively, not quantitatively. In addition, the concept of mean free path is different in different problems of physical kinetics.

The Boltzmann kinetic equation determines the microscopic evolution of the state of a nonequilibrium gas. In the derivation of this equation one assumes that molecular collisions have zero time duration and occur at any point of space. Therefore the distribution function obtained from the solution of the Boltzmann equation is valid for time durations which are large in comparison with the time duration of the collisions themselves. In addition, the distances are assumed to be large in comparison with the dimensions of the collision region, i.e., the range of the intermolecular forces.

In the qualitative treatment of problems of physical kinetics we used the concept of a weakly nonuniform gas. In the language of the Boltzmann equation this means that the deviation of the distribution function from the equilibrium distribution function is small. In this approximation the collision integral can be simplified and the Boltzmann equation can be linearized in the correction to the equilibrium distribution function. As a result, the Boltzmann equation is simplified considerably, but remains an integrodifferential equation.

In the simplification described above, the collision integral has the equilibrium distribution function (the Maxwellian velocity distribution) under the integral sign. Recall that the validity of the Maxwellian distribution does not require that the gas be ideal.

Even the linearized Boltzmann distribution is not so easy to solve, since it is still an integral equation. A general method of solution is to expand the correction to the equilibrium distribution function in a complete set of orthogonal functions. These functions are chosen such that their orthogonality can be used effectively in obtaining equations for the expansion coefficients. As noted above, the orthogonality condition must contain the equilibrium distribution function, i.e., the exponential in the Maxwellian distribution. Hence the orthogonal functions must be such that the weight function in the orthogonality condition is an exponential. As is known from mathematical physics, such functions are the *generalized Laguerre polynomials.* In the kinetic theory of gases one usually uses the so-called *Sonine polynomials,* which differ from the generalized Laguerre polynomials only by a normalization factor.

The convergence of the approximation process depends on which kinetic coefficient is being calculated. The series expansion does not contain a small parameter following from physical considerations. The rapidity of the convergence depends solely on the smallness of the terms of the series.

The coefficients of the expansion in Sonine polynomials are represented by integrals containing the differential cross section for scattering of molecules by one another. Therefore the explicit form of these expansion coefficients can only be obtained after an explicit expression for the cross section is substituted in the integrals. For example, we can assume that the molecules behave as hard elastic spheres and calculate the expansion coefficients in this approximation. The expansion is truncated after a given number of terms, depending on the accuracy required of the solution. The kinetic coefficients of a given physical problem are calculated in this way.

From the above discussion one can imagine the extreme complexity of numerical solutions of problems in physical kinetics. For this reason qualitative estimates are extremely useful, in spite of the fact that they cannot predict numerical factors.

The macroscopic equations of hydrodynamics are obtained from the Boltzmann kinetic equation by averaging over the variables on which the distribution function depends. This averaging is valid when the distance over which the macroscopic parameters of the medium vary is large in comparison with the mean free path of the molecules. Hence the macroscopic description of the medium assumes that the gradients of the macroscopic parameters are sufficiently small. This fact was used in many sections of this book to obtain a phenomenological representation of the physical quantities of interest.

For example, from the point of view of macroscopic gas dynamics, the width of a strong shock wave must be zero and the methods of gas dynamics are incorrect in studying the structure of the front of a shock wave: they are valid only for weak shock waves or for a medium with large viscosity or thermal conductivity. In such a medium nonlinear effects lead to a steepening of the wave front with time, which may lead to discontinuities in the hydrodynamic parameters typical of shock waves. However, the growth of the gradients of the hydrodynamic quantities amplifies dissipative effects, which are proportional to these gradients. Dissipative effects, on the other hand, decrease the steepness of the profile of the wave front. Competition between these two effects determines the width of the zone over which the discontinuities in the hydrodynamic parameters occur. The applicability of the gas-dynamical approach to the structure of the front depends on this width.

The basic difficulty of the hydrodynamic equations lies in the nonlinear term $(\mathbf{v}\cdot\nabla)\mathbf{v}$ in the Navier–Stokes equation. We saw in Sec. 8.4 that this nonlinearity leads to nontrivial effects: period doubling of oscillating solutions was an example.

Furthermore, according to the qualitative approach, the transition from laminar flow to turbulent flow should occur at Reynolds numbers of order unity (or at least not too different from unity). However, for flow in pipes it is known experimentally that the critical Reynolds number is several thousand. Obviously, the qualitative approach cannot explain where such a large value comes from. For a linear differential equation with coefficients of order unity and subject to an initial condition on a dimensionless combination of the physical quantities of order unity, the dimensionless solution of the equation must also be of order unity. We have seen this many times in different problems of physical kinetics and hydrodynamics.

Hence the important effects are connected with the nonlinearity of the Navier–Stokes equation. However, in the nonlinear example considered in Sec. 8.4, the critical Reynolds number turned out to be 1.37. It is possible that in real three-dimensional problems of turbulent flow the mechanism determining the onset of turbulence involves an exponential dependence of the critical Reynolds number on the parameters of the flow (or a power law with a large exponent).

It follows that the qualitative method requires the use of experimental data. This is because the solutions of the nonlinear Navier–Stokes equation are extremely complicated and the basic features of the solution (which may not be complicated in themselves) cannot be predicted using only the form of the equation. Exact solutions for three-dimensional flow fields in realistic situations are scarce, despite the availability of powerful computers. For example, the solution presented in Sec. 8.4 is only one of the mechanisms determining the onset of turbulence.

The nonlinearity of the hydrodynamic equations can also lead to other very interesting effects, which are outside the scope of the present text. An example is the so-called *soliton solutions*. Solitons are solitary waves which propagate without changing their shape. They decay in both directions at infinity. They exist in media without dissipation. As in the case of formation of shock waves, nonlinear effects lead to a gradual steepening of the leading front of the wave. The wave profile spreads out because of dispersion in the medium, rather than because of dissipation. These two effects can compensate one another and the result is that the profile of the soliton remains constant in time.

Finally, we note that in contrast to systems in statistical equilibrium, kinetic processes are intimately connected with the nature of the microscopic interactions between particles; this is why there is such a great diversity of phenomena. This book includes only a small fraction of such problems. The examples were chosen to illustrate typical qualitative methods of kinetic theory and hydrodynamics.

Recommended literature

Chapters 1–5

Ashcroft, N. W., and Mermin, N. D., *Solid State Physics,* 4th ed. (Holt, Rinehart and Winston, New York, 1976).
Aubry, S., *Soliton and Condensed Matter Physics,* edited by A. Bishop and T. Schneider (Springer, New York, 1979), p. 264.
Brittin, W. (ed.), *Lectures in Theoretical Physics,* Vol. IX C, Kinetic Theory (Gordon and Breech, New York, 1967).
Chapman, C., and Cowling, T. G., *Mathematical Theory of Non-Uniform Gases,* 3rd ed. (Cambridge University Press, Cambridge, 1970).
Cohen, E. G. D., *On the Statistical Mechanics of Moderately Dense Gases not in Equilibrium* (University of Colorado, Boulder, 1966).
Elliott, R. J., and Gibson, A. F., *Solid State Physics and Its Applications* (Harper and Row and Barnes and Noble Import Division, New York, 1974).
Ferziger, J. H., and Kaper, H. G., *Mathematical Theory of Transport Process in Gases* (North-Holland, Amsterdam, 1972).
Grad, H., *Principles of the Kinetic Theory of Gases,* in Handbuch der Physik, Band 12 (Springer, Heidelberg, 1958), p. 205.
Hirschfelder, J. O., Curtiss, C. F., and Bird, R. B., *Molecular Theory of Gases and Liquids* (Wiley, New York, 1954).
Kittel, C., *Introduction to Solid State Physics,* 5th ed. (Wiley, New York, 1976).
Kogan, M. N., *Rarefied Gas Dynamics* (Plenum, New York, 1969).
Lifshitz, E. M., and Pitaevskii, L. P., *Physical Kinetics* (Pergamon, Oxford, 1981).
Reid, R. C., Prausnitz, J. M., and Sherwood, T. K., *The Properties of Gases and Liquids* (McGraw-Hill, New York, 1977).
Schaaf, S. A., and Chambré, P. L., *Flow of Rarefied Gases* (Princeton University Press, Princeton, 1961).
Vincenti, W. G., and Kruger, C. H., Jr., *Introduction to Physical Gas Dynamics* (Wiley, New York, 1965).
Ziman, J. M., *Principles of the Theory of Solids* (Cambridge University Press, London, 1964).

Chapters 6–13

Astarita, G., and Marucci, G., *Principles of Non-Newtonian Fluid Mechanics* (McGraw-Hill, London, 1974).
Bradshaw, P., *Experimental Fluid Mechanics* (Pergamon, New York, 1964).
Bradshaw, P., *Introduction to Turbulence and Its Measurement* (Pergamon, New York, 1971).
Chandrasekhar, S., *Hydrodynamic and Hydromagnetic Stability* (Clarendon, Oxford, 1961).
Feigenbaum, M., J. Stat. Phys. **19**, 25 (1978) (to Sec. 8.4).
Hinze, J. O., *Turbulence,* 2nd ed. (McGraw-Hill, New York, 1975).

Lamb, H., *Hydrodynamics,* 6th ed. (Cambridge University Press, London, 1932).
Landau, L. D., and Lifshitz, E. M., *Fluid Mechanics,* 2nd ed. (Pergamon, London, 1984).
Lichtenberg, A. J., and Liberman, M. A., *Regular and Stochastic Motion* (Springer, New York, 1983).
Oswatitsch, K., *Grundlagen der Gasdynamik* (Springer, Wien, 1976).
Rouse, H., and Howe, J. W., *Basic Mechanics of Fluids* (Wiley, New York, 1953).
Schlichting, H., *Boundary-Layer Theory,* 7th ed. (McGraw-Hill, New York, 1979).
Shames, I. H., *Mechanics of Fluids* (McGraw-Hill, New York, 1962).
Streeter, V. L. (ed.), *Handbook of Fluid Dynamics* (McGraw-Hill, New York, 1961).
Swinney, H. L., and Golub, J. P., *Hydrodynamics, Instabilities, and the Transition to Turbulence* (Springer, Berlin, 1981).
Tennekes, H., and Lumley, J. L., *A First Course in Turbulence* (MIT, Cambridge, 1974).
Tritton, D. H., *Physical Fluid Dynamics* (Van Nostrand Reinhold, New York, 1977).

Notation

a	Acceleration of a body, oscillation amplitude, thermal diffusivity
a_B	Bohr radius
A	Work
α	Recombination coefficient, thermoelectric coefficient, surface tension, coefficient of linear expansion
α_M	Mach angle
α_r	Photorecombination coefficient
α_s	Absorption coefficient of sound
b	Width of a body
β	Coefficient of volume expansion
c	Concentration, speed of light
c_c	Velocity of capillary waves
c_g	Velocity of gravity waves
c_i	Concentration of ions
c_p	Phase velocity of light in the plasma
c_P	Specific heat at constant pressure
c_s	Speed of sound
C	Heat capacity per unit volume of material
γ	Adiabatic exponent, damping coefficient
d	Molecular diameter, lattice constant
D	Diffusion coefficient, hydraulic diameter
D_a	Coefficient for ambipolar diffusion
D_e	Diffusion coefficient of the electrons
D_i	Diffusion coefficient of the ions
D_{td}	Thermal diffusion coefficient
$\mathbf{D}$	Electric displacement
δ	Thickness of the boundary layer
Δ	Difference, Laplacian operator
e	Charge of the electron
E	Energy, thermal energy per molecule, electric field strength
$\dot{E}$	Dissipated power
E_D	Dissociation energy
ε	Energy per particle, dielectric permittivity, porosity, power absorbed per unit mass of the fluid
ε_0	Permittivity of free space
F	Force
g	Acceleration of gravity, temperature jump coefficient
h	Height, depth, heat transfer coefficient
$\hbar$	Planck's constant
H	Height, depth
η	Dynamical viscosity
i	Diffusion flux density of particles

I	Ionization potential, moment inertia of a molecule, intensity of acoustic radiation
j	Current density, dissipated power per unit volume
k	Photon momentum, wave number, permeability constant
K	Knudsen number
l	Mean free path, length of a pipe, scale of turbulent flow
l_0	Minimum scale of turbulent flow
L	Path length, Coulomb logarithm, total angular momentum
λ	Wavelength, thermal conductivity
λ_{De}	Debye length for electrons
λ_{Di}	Debye length for ions
m	Mass of the electron, mass of the drop
M	Mass of a molecule, mass of a particle
Ma	Mach number
μ	Reduced mass of particles, mobility
μ_t	Thermal slip coefficient
n	Concentration of molecules
n_e	Concentration of electrons
n_i	Concentration of ions
n_a	Concentration of neutral molecules
N	Number of collisions
Nu	Nusselt number
ν	Kinematic viscosity, collision frequency of molecules
ν_e	Collision frequency of electrons
$\mathbf{p}$	Momentum of a particle
P	Pressure
$\mathbf{P}$	Polarization
Pr	Prandtl number
Π	Momentum flux of matter
q	Flux density of thermal energy, surface flux density of thermal energy
Q	Mass flow rate through a pipe, volume energy density
r	Distance, specific heat of vaporization
$\langle r \rangle$	Average distance between molecules in a gas
R	Linear dimension of a body, pipe radius
Re	Reynolds number
Re_{cr}	Critical Reynolds number
$\mathscr{R}$	Rayleigh number
ρ	Density, impact parameter of a collision, electrical resistivity
S	Surface area
σ	Collision cross section of molecules, cross section for scattering of sound
σ'	Cross section for absorption of sound
σ_e	Electrical conductivity
σ_i	Ionic conductivity, cross section for ionization
σ_r	Cross section for recombination
σ_t	Cross section for momentum transfer in scattering
t	Time
T	Temperature, period of the oscillations
T_e	Electron temperature
T_i	Ion temperature
τ	Mean free time, relaxation time
τ_c	Collision time
τ_D	Diffusion time
Θ	Debye temperature

θ	Scattering angle
u	Velocity of particles, stream velocity of a fluid
U	Potential energy
v	Velocity of particles
v_t	Thermal velocity
V	Drift velocity of particles, beam velocity, volume of a system
x,y,z	Coordinates of a point
χ	Electric susceptibility
W	Energy flux density of matter
ω	Wave frequency
ω_{pe}	Electronic plasma frequency
ω_{pi}	Ionic plasma frequency

Printed in the United States
By Bookmasters